# SQL Simplified:
## Learn To Read and Write Structured Query Language

By

Cecelia L. Allison

© 2004 by Cecelia L. Allison. All rights reserved.

No part of this book may be reproduced, stored in a retrieval system, or transmitted by any means, electronic, mechanical, photocopying, recording, or otherwise, without written permission from the author.

ISBN: 1-4107-2973-7 (e-book)
ISBN: 1-4107-2974-5 (Paperback)

Library of Congress Control Number: 2003091775

This book is printed on acid free paper.

Printed in the United States of America
Bloomington, IN

Editor: John F. Allison

Development Editor: Panzina Elizabeth Coney

1stBooks – rev. 01/07/04

# Contents

About the Author .................................................. ix
Dedication ............................................................ xi
Acknowledgments ............................................. xiii
Introduction ........................................................ xv
    Platforms used in the Book .......................... xv

**Chapter 1. The Relational Database and Structured Query Language .................................. 1**
    The Relational Database Structure ................. 2
    Structured Query Language ........................... 5

**Chapter 2. Creating Tables and Working with Data ..................................................................... 8**
    Table Creation ................................................. 8
    Inserting Data Into a Table ........................... 15
    Transferring Data from One Table to an Existing Table .............................................. 19
    Transferring Data from One Table to a New Table ..................................................... 20
    Updating Data in a Table .............................. 21
    Deleting Data in a Table ............................... 23

**Chapter 3. Selecting and Retrieving Data ......... 25**
    The SELECT Statement ............................... 26

Creating a SELECT Statement to Retrieve Multiple Columns ................................................. 27
Creating a SELECT Statement to Retrieve Every Column ................................................. 28
Using DISTINCT to display Unique Values in a Column ................................................. 29
Using the AS Keyword to Create an Alternate Name for a Column ................................................. 30
Merging Columns ................................................. 32
Uniting Queries to Compare Records in Two Separate Tables ................................................. 33

## Chapter 4. Filter Retrieved Data ................................................. 38

The WHERE Clause ................................................. 38
Using a Comparison Operator to Match a Condition ................................................. 40
Using a Character Operator to Match a Condition ................................................. 42
Using a Logical Operator to Match a Condition ................................................. 45
Using a Logical Operator to Match a Condition Opposite of the One Defined ....... 46
Using the IN Operator to Match a Condition ................................................. 47
Using the BETWEEN Operator to Match a Condition ................................................. 49

## Chapter 5. Creating Calculated Fields .............. 52
Calculated Fields ............................................. 52
Concatenation .................................................. 53
Arithmetic Operators ....................................... 54
Functions ......................................................... 56

## Chapter 6. Using Additional Clauses in Structured Query Language .............................. 66
Clauses ............................................................. 67
Using the ORDER BY Clause ....................... 67
Using the GROUP BY Clause ....................... 74
Using the GROUP BY and WHERE Clause ............................................................. 76
Using the HAVING Clause ........................... 77

## Chapter 7. Creating Table Joins ....................... 81
Table Joins ....................................................... 82
Qualification .................................................... 82
Self Join ........................................................... 86
Natural Join ..................................................... 89
Outer Join ........................................................ 92

## Chapter 8. Creating Subqueries ....................... 100
Subqueries ....................................................... 101
Using the IN keyword to link Queries ........... 102
Using the EXISTS keyword to link Queries ............................................................ 104
Using the ANY keyword to link Queries ... 105

Using the ALL keyword to link Queries..... 107
Creating Subqueries to Query Three Tables 109
Creating Subqueries that Contain
Aggregate Functions.................................. 112

## Chapter 9. Creating Views.................................117

Views .............................................................. 118
Creating a View on One Table ................... 118
Creating a View on Multiple Tables........... 121
Filtering Table Data Using a View ............. 124
Updating and Inserting Data into Tables
Using Views.................................................. 125
Deleting a View ............................................ 129

## Chapter 10. Advanced Table Creation and Table Management..................................................131

Table Creation .............................................. 132
Constraints .................................................... 134
Indexes.......................................................... 136
Table Management ...................................... 139
Adding a Column......................................... 139
Deleting a Column....................................... 141
Deleting a Table........................................... 141

## Chapter 11. Advanced Topics: Transaction Processing, Stored Procedures, Triggers and Cursors ...............................................................143

Transaction Processing ............................... 145
Stored Procedures ........................................ 148

Triggers ......................................................... 155
　　　Cursors .......................................................... 158
**Appendix A** ........................................................... 162
　　　Answers to Quizzes and Assignments ........ 162
**Appendix B** ........................................................... 171
　　　SQL Script for the Tables Used in the
　　　Book ............................................................... 171
**Appendix C** ........................................................... 187
　　　SQL Command Syntax .............................. 187
**Appendix D** ........................................................... 190
　　　Instructions on Where to Type SQL Script
　　　in Microsoft Access and Microsoft SQL
　　　Server ........................................................ 190
**Index** ....................................................................... 193

# About the Author

Cecelia L. Allison is originally from St. Petersburg, Florida. She is a freelance Webmaster and also works as an author/facilitator instructing students on Structured Query Language (SQL) over the Internet (www.jaffainc.com) through universities, colleges, and other educational facilities around the world.

Cecelia has been using SQL for many years. Through her past and present employment she has gained extensive experience in writing and implementing SQL. She holds a Bachelor of Science degree in Finance and a Master of Science degree in Computer Information Systems.

# Dedication

This book is dedicated to my daughter Kayla Allison and my husband John F. Allison. John you inspired, supported, loved, and encouraged me throughout this project. You are truly a blessing to my life. Thank you for all of your support and assistance. I love you dearly.

# Acknowledgments

Thank you **Panzina E. Coney** for all of your hard work in making this book possible. Your contribution is greatly appreciated.

I want to thank my parents **Rosa D. Coney and Willie A. Coney**. Thank you for raising me to always give my best at whatever I do and for instilling exceptional values and work ethics that will last a lifetime. I love you.

Thank you **Teik-Seng Yu, aka Cowboy**, for all of your support. Your kindness and patience meant a lot.

Thank you **Richard and Gayle Finch** for taking me under your wing and inspiring me to write and publish this book.

I'd also like to thank Dion and Stephanie Dixon, Thomas and Debra Brown, Leonard and Yolanda Cole, Vernon and Yvonne Spellman, Dexter and Tanya Levin, Otis Coney, Yolanda Love, and LaShawn Jackson.

# Introduction

Structured Query Language (SQL) is a database programming language used to create and maintain relational databases. It is used to create tables, sort data, retrieve data, update data and to set security settings.

SQL is the industry standard database programming language. This book is designed to provide beginners with the tools essential for learning how to read and write SQL.

Instead of focusing primarily on database design or on the Database Management Systems (DBMSs) that implement SQL, like many SQL books; this book focuses extensively on the implementation of SQL. Whether for business or home, the easy to follow step-by-step chapters of this book will give beginners the practice necessary to develop the skills and knowledge necessary to program in SQL with ease.

The concepts of SQL are simplified enabling anyone to quickly grasp the fundamentals of SQL. Each chapter introduces a new concept and includes examples, key notes and important key terms. Your comprehension of each chapter is tested through the use of quizzes and assignments. After completion of this book, you should feel confident using SQL in any relational database environment.

## Platforms used in the Book

The examples in this book were created using Microsoft SQL Server and Microsoft Access. Since most SQL script can be transferred to other Relational Database Management Systems (RDBMS) with very little modification, you can use other DBMSs to practice the examples used in this book.

Some script modifications for other DBMSs are covered in this book. Refer to your DBMS documentation for script modifications not covered. Appendix B contains the necessary script to create the tables used throughout the book.

# Chapter 1

# The Relational Database and Structured Query Language

## Introduction

In this chapter, you will learn about the structure and creation of relational database structures. You will also learn about Structured Query Language (SQL), the database programming language used to manage relational database structures.

Structured Query Language (SQL) is widely used by many industry professionals. It is transferable among several database platforms and is very easy to learn.

Before diving right into Structured Query Language, you need a basic understanding of the relational database structure. Read over the important terms and concepts below.

## Important Terms:

**Column**: A separate entity within a record that runs vertically within a table.

**Foreign Key**: A foreign key links records of one table to those in another table.

**Keys**: Used to uniquely identify a row or record within a table.

**Keyword**: Reserved words that are used to interact with the database.

**Normalization**: A technique used to organize data attributes in a more efficient, reliable, flexible and maintainable structure.

*Chapter 1*

**Primary Key**: A field whose value uniquely identifies every row in a table.

**Query**: A request or command to the DBMS.

**Relational Database**: A relational database is a collection of two or more two-dimensional tables related by key values.

**Row**: A record within a table that runs horizontally within a table.

**Structured Query Language**: A database programming language used to create tables in a relational database; sort, retrieve, and update data stored in a relational database and set security settings.

**Syntax**: Rules you must follow when writing SQL script.

**Table**: A two-dimensional file containing rows and columns.

# The Relational Database Structure

A *relational database* is a collection of two or more two-dimensional tables related by key values. The *tables* in a relational database are two-dimensional because they contain columns that run vertically, and rows that run horizontally. Each *row* in a table indicates the total number of records in a table, and each *column* indicates a separate entity within a record. Look at figure 1.1.

## Members Table

| MemberID | Firstname | Lastname | Address | City | State | Zipcode | Areacode | PhoneNumber |
|---|---|---|---|---|---|---|---|---|
| 1 | Jeffrey | Lindley | 3980 14th Ave S | Atlanta | GA | 98700 | 301 | 451-5451 |
| 2 | Jerry | Lindsey | 4000 3rd Ave S | Tampa | FL | 33600 | 813 | 923-7852 |
| 3 | Gerry | Pitts | 3090 13th St N | Tampa | FL | 33611 | 813 | 286-4821 |
| 4 | Stan | Benson | 1825 8th St N | Santa Fe | NM | 88388 | 505 | 464-1578 |
| 5 | Peter | Gable | 1097 10th Ave S | St. Petersburg | FL | 33754 | 727 | 327-1253 |

**Figure 1.1**

*The Relational Database and Structured Query Language*

Figure 1.1 illustrates a table named Members with five records (rows) and each record contains nine separate entities (columns) of information (MemberID, Firstname, Lastname, Address, City, State, Zipcode, Areacode and PhoneNumber).

# Keys

To relate tables to one another in a relational database structure you must use keys. *Keys* are used to uniquely identify a record within a table. The primary key and the foreign key are the two most commonly used keys during table creation.

A *primary key* is a field or column whose value uniquely identifies every row in a table. Once you know what columns you want to include in a table, you must decide which column to designate as the primary key column. In Figure 1.2, the StudentID column is the primary key column because it contains only unique numbers. No two students can have the same student ID number.

A *foreign key* links records of one table to those of another table. Specifically, a foreign key points to records of a different table in a database. Look at figure 1.2 and 1.3.

## Students Table

| StudentID | Name | Address | Zipcode | PhoneNumber |
|---|---|---|---|---|
| 1 | John Watkins | 1429 24th Ave N St. Petersburg, FL | 33711 | (727)327-8428 |
| 2 | Jason Little | 2222 5th Ave N St. Petersburg, FL | 33711 | (727)323-3490 |
| 3 | Rosa Coney | 1601 First St N Tampa, FL | 33612 | (813)332-2635 |
| 4 | John F. Allison | 114-C 20th St N Honolulu, HI | 42319 | (808)239-2929 |
| 5 | Debra Brown | 1934 Second St. N Tampa, FL | 33619 | (813)253-0977 |

## Figure 1.2

## Courses Table

| CourseID | StudentID | Course | StartTime | EndTime | StartDate | EndDate | Teacher | Credit |
|---|---|---|---|---|---|---|---|---|
| L1001 | 3 | Literature | 2:00pm | 4:00pm | 2/3/2003 | 5/3/2003 | Mrs. Donaldson | 3 |
| M1101 | 1 | Pre Algebra | 3:00pm | 5:00pm | 2/3/2003 | 5/3/2003 | Mr. Stevens | 3 |
| M1102 | 5 | Pre Calculus | 3:00pm | 5:00pm | 2/3/2003 | 5/3/2003 | Mr. Dixon | 3 |
| M1103 | 4 | Statistics | 3:00pm | 5:00pm | 2/3/2003 | 5/3/2003 | Mr. Levin | 3 |
| R1001 | 2 | Reading | 1:00pm | 3:00pm | 2/3/2003 | 5/3/2003 | Ms Jackson | 3 |

**Figure 1.3**

In the Courses table, the StudentID column is a foreign key because the StudentID column links the student's records from the Students table to the records stored in the Courses table. As you can see, the StudentID column is a primary key in the Students table and a foreign key in the Courses table. To create a foreign key, you must create a primary key column in one table and duplicate that column in another table.

## Database Planning Phase

Before you create tables for your database, first carefully plan and outline the tables, columns, and keys. Careful planning in the beginning will save you valuable time in the future.

Begin by asking yourself or the clients why the system is needed, what the database will track, and who will use the system. Create your tables based on this information.

A widely excepted technique that is used in the creation of databases is called normalization. *Normalization* is a technique used to organize data attributes in a more efficient, reliable, flexible and maintainable structure. There are three phases in the normalization process.

Phase one is called first normal form. In first normal form, you eliminate repeating groups of information and create primary keys. Tables that contain additional groups of information are broken down into separate tables and each table is assigned a primary key. Each table should only contain one group of information. For example, a table with information about students should not include information about another group, such as teachers.

Phase two is called second normal form. In second normal form, make sure every column in the table is dependent on the primary key. Continue to break your tables into smaller ones if possible. Create as many tables as needed so that each table contains a separate group of information.

Phase three is called third normal form. In third normal form, totally remove columns that create a separate group of information and are not stored in a separate table.

Now that you have a basic understanding of the relational database structure, let's discuss Structured Query Language.

## Structured Query Language

*Structured Query Language* (SQL) is a database programming language used to create tables in a relational database; sort, retrieve, and update data stored in a relational database and set security settings.

### History

IBM developed SQL in 1970, which means it's been around for over 30 years. Its original name was SEQUEL. SEQUEL stands for Structured English Query Language.

In 1979, a company called Relational Software, Inc., implemented SQL. Today, Relational Software, Inc is known as the Oracle Corporation.

*Chapter 1*

## Today

Currently, SQL is widely implemented by many companies in the industry and almost all major relational database management systems support SQL. Some relational database management systems use slightly different rules for writing SQL however, most of the script written in one RDBMS can be moved into another with minimum modification.

Some of the most widely used relational database management systems include, Microsoft Access, Microsoft SQL Server, Oracle, MySQL, and DB2.

## SQL Rules

In order to use Structured Query Language, you must locate the area in which to type SQL script in a DBMS. In Microsoft SQL Server, this area is called Query Analyzer. In Microsoft Access it is called Query Designer. Refer to Appendix D for instructions on where to type SQL script in Microsoft SQL Server and Microsoft Access.

SQL enables you to communicate with the DBMS. You use a combination of keywords that the DBMS understands. In chapter two, you will learn some of the keywords used in SQL. *Keywords* are reserved strictly for interacting with the database therefore; you cannot use any of the keywords to name tables, columns, or other portions of the database. You also must use the correct syntax to avoid system-generated errors. *Syntax* refers to the rules you must follow when writing SQL script. You will learn more on syntax in chapter two.

# Conclusion

In this chapter, you learned important key terms and about the structure of relational database structures. You also learned about the birth of Structured Query Language (SQL), and its current impact on relational database structures.

# Test Your Knowledge of the Chapter

# Quiz 1

1. True or False: A column is a record within a table that runs horizontally within a table.
2. True or False: Microsoft created Structured Query Language.
3. True or False: A primary key is a field whose value uniquely identifies every row in a table.
4. True or False: Normalization is a technique used to organize data attributes in a more efficient, reliable, flexible and maintainable structure.
5. What links records of one type to those of another type?

# Assignment 1

On paper, create columns for two tables. Link the two tables by assigning a primary key to both tables and a foreign key to one table.

# Chapter 2

# Creating Tables and Working with Data

## Introduction

In this chapter, you will learn how to use Structured Query Language (SQL) to create a table and populate it with data. You will also learn how to alter data stored in a table.

## Table Creation

Structured Query Language (SQL) is primarily used to query databases, but you can also use SQL to create and populate tables stored in a relational database structure.

Take a look at the important terms for this chapter.

## Important Terms:

**Asterisk (*)**: A symbol often used with the SELECT keyword to tell the DBMS to select every column.

**CREATE TABLE**: Keywords used to create a new table.

**Datatype**: Specifies the type of data a column can store.

**DELETE**: Used to remove records from a table.

**Field Size**: Specifies the maximum number of characters that can be entered into a cell within a column.

**Field**: Equivalent to a column.

*Creating Tables and Working with Data*

**FROM**: Keyword used to specify specific table(s) in a database.

**INSERT Statement**: Used to insert a single row into a table.

**NOT NULL**: Indicates that a field cannot be left blank when entering data into a table.

**NULL**: Indicates that a field can be left blank when entering data into a table.

**SELECT**: Keyword used to specify specific column(s) from a table.

**SET**: Used to assign a new value to a field.

**UPDATE**: Used to update one or more rows in a table.

**WHERE**: Used to set a condition in a query.

 The examples in this chapter were created using Microsoft SQL Server.

Look at the SQL syntax to create a table.

## Syntax

### CREATE TABLE Syntax

CREATE TABLE TableName
(
ColumnOne Datatype [NULL | NOT NULL],
ColumnTwo Datatype [NULL | NOT NULL],
ColumnThree Datatype [NULL | NOT NULL]
);

*Chapter 2*

The CREATE TABLE syntax shows the proper format to follow to create a new table. It enables you to create a table name, column names, datatypes etc. Example one shows how to create a table.

 Some Database Management Systems (DBMSs) may use slightly different syntax; check your DBMS documentation for changes.

## CREATE TABLE Employees

## Example 1

The following SQL script creates a table named Employees with seven columns (SocialSecNum, Firstname, Lastname, Address, Zipcode, Areacode, PhoneNumber). Figure 2.1 shows the Employees table created using the CREATE TABLE script.

```
CREATE TABLE Employees
(
SocialSecNum CHAR (11) NOT NULL PRIMARY KEY,
Firstname CHAR (50) NOT NULL,
Lastname CHAR (50) NOT NULL,
Address CHAR (50) NOT NULL,
Zipcode CHAR (10) NOT NULL,
Areacode CHAR (3) NULL,
PhoneNumber CHAR (8) NULL
);
```

The preceding script includes several keywords. The keywords are typed in all caps so that they can be easily identified. Although it is not necessary to capitalize keywords, it is good practice since it causes the keywords to stand out. This makes SQL script easier to read. Remember, *keywords* are reserved words used to interact with the database.

*Creating Tables and Working with Data*

Notice the spacing in the CREATE TABLE script. You must separate all keywords and column names with a space. Additionally, some DBMSs require a closing semicolon at the end of SQL script to indicate where the script ends.

The *CREATE TABLE* keywords tell the database management system that you want to create a new table. The name (Employees) of the table must be typed directly after the CREATE TABLE keywords.

Each column (SocialSecNum, Firstname, Lastname, Address, Zipcode, Areacode and PhoneNumber) must contain a datatype and a field size. Some datatypes do not require a field size.

A *datatype* specifies the type of data a column can store. The *field size* specifies the maximum number of characters that can be entered into a cell within a column. For example, the datatype and field size for the first column (SocialSecNum) is CHAR (11). This means that the SocialSecNum column can only store alphanumeric data no more than eleven characters long.

Following the datatype and field size, specify either NULL or NOT NULL. The *NULL* keyword indicates that a column can be left blank when entering data in the table. The *NOT NULL* keywords indicate that a column cannot be left blank. The DBMS will generate an error message whenever a NOT NULL field is left blank.

In Microsoft Access, when you do not state NULL or NOT NULL the column is automatically set to NULL.

The primary key is usually specified after NULL | NOT NULL. In figure 2.1, the SocialSecNum column is the primary key column. In some versions of Microsoft Access, the primary key is set as follows:

*Chapter 2*

SocialSecNum CHAR (11) NOT NULL CONSTRAINT PriKey Primary Key

This method uses the CONSTRAINT keyword. You will learn more about constraints in chapter 10.

Figure 2.1 shows the Employees table created using the CREATE TABLE script.

## Employees Table

| SocialSecNum | Firstname | Lastname | Address | Zipcode | Areacode | PhoneNumber |
|---|---|---|---|---|---|---|
| | | | | | | |

**Figure 2.1**

## Datatypes

Below are some of the most commonly used datatypes, supported by Microsoft Access and Microsoft SQL Server.

## Microsoft Access

**Text**: Variable length datatype. Stores up to 255 characters of text and numbers.

**Counter**: Stores long integer values that automatically increment whenever a new record is inserted into a table.

**Char**: Fixed length datatype. Stores alphanumeric data up to 255 characters.

**Binary**: Stores any type of data.

**DateTime**: Stores date and time values.

**Bit**: (Yes/No data type) Stores yes and no, true and false, and on and off values.

**Money**: Stores currency values and numeric data used in mathematical calculations.

**Real**: Stores single-precision floating-point values.

**Image**: Stores objects such as Microsoft Excel spreadsheets, Microsoft Word documents, graphics, sounds, or other binary data linked to or embedded in a table.

**Int**: Stores a long integer between −2,147,483,648 and 2,147,483,647.

**Float**: Stores double-precision floating-point values.

*Chapter 2*

# Microsoft SQL Server

**Decimal**: Fixed precision and scale numeric data

**Numeric**: A synonym for decimal.

**Money**: Stores currency values. Monetary data values vary.

**Smallmoney**: Stores currency values. Monetary data values vary.

**Real**: Floating precision number data from -3.40E + 38 through 3.40E + 38.

**Float**: Floating precision number data from -1.79E + 308 through 1.79E + 308.

**Int**: Integer (whole number) data between -2,147,483,648 and 2,147,483,647.

**Bit**: Integer data with either a 1 or 0 value.

**Smallint**: Integer data between -32,768 and 32,767.

**Tinyint**: Integer data from 0 through 255.

**Datetime**: Date and time data from 01/01/1753 to 12/31/9999

**Smalldatetime**: Date and time data from 01/01/1900, through 06/06/2079

**Timestamp**: A database-wide unique number.

**Uniqueidentifier**: A globally unique identifier.

**Image**: Variable-length binary data with a maximum length of 2^31 - 1 (2,147,483,647) bytes.

**Binary**: Fixed-length binary data with a maximum length of 8,000 bytes.

**Varbinary**: Variable-length binary data with a maximum length of 8,000 bytes.

**Cursor**: A reference to a cursor.

**Char**: Fixed-length non-Unicode character data with a maximum length of 8,000 characters.

**Varchar**: Variable-length non-Unicode data with a maximum of 8,000 characters.

**Text**: Variable-length non-Unicode data with a maximum length of $2^{31} - 1$ (2,147,483,647) characters.

**Nchar**: Fixed-length Unicode data with a maximum length of 4,000 characters.

**Nvarchar**: Variable-length Unicode data with a maximum length of 4,000 characters.

**Ntext**: Variable-length Unicode data with a maximum length of $2^{30} - 1$ (1,073,741,823) characters.

---

**Table 2.1.** Microsoft Access and Microsoft SQL Server Datatypes

# Inserting Data Into a Table

## Populate the Employees Table

# Example 2

Figure 2.2 shows the Employees table created in example one. To populate the Employees table using SQL, you must create insert statements.

## Employees Table

| SocialSecNum | Firstname | Lastname | Address | Zipcode | Areacode | PhoneNumber |
|---|---|---|---|---|---|---|
| | | | | | | |

**Figure 2.2**

Look at the SQL syntax to insert a single row (record) into a table.

## INSERT Syntax

INSERT INTO TableName [(ColumnNames, ...)]
VALUES (values, ...);

The INSERT syntax shows the proper format to follow to insert a new record into a table. It enables you to insert column values into a specified table. Look at the following insert statements that insert five records into the Employees table.

 Some DBMSs may use slightly different syntax; check your DBMS documentation for changes.

## INSERT Statements

INSERT INTO Employees (SocialSecNum, Firstname, Lastname, Address, Zipcode, Areacode, PhoneNumber)
VALUES ('266-73-1982', 'John', 'Dentins', '2211 22nd Ave N GA, Atlanta', 98718, 301, '897-4321');

INSERT INTO Employees (SocialSecNum, Firstname, Lastname, Address, Zipcode, Areacode, PhoneNumber)
VALUES ('266-11-4444', 'Sam', 'Elliot', '1601 Center Loop Tampa, FL', 33612, 813, '898-2134');

*Creating Tables and Working with Data*

INSERT INTO Employees (SocialSecNum, Firstname, Lastname, Address, Zipcode, Areacode, PhoneNumber)
VALUES ('263-73-1442', 'Adam', 'Williams', '1938 32$^{nd}$ Ave S. St. Pete, FL', 33711, 727, '321-2234');

INSERT INTO Employees (SocialSecNum, Firstname, Lastname, Address, Zipcode, Areacode, PhoneNumber)
VALUES ('226-73-1919', 'Jacob', 'Lincoln', '2609 40th Ave S HI, Honolulu', 96820, 808, '423-4111');

INSERT INTO Employees (SocialSecNum, Firstname, Lastname, Address, Zipcode, Areacode, PhoneNumber)
VALUES ('249-74-1682', 'Jackie', 'Fields', '2211 Peachtree St N Tampa, FL', 33612, 813, '827-2301');

In Microsoft SQL Server, you can run all five INSERT statements at the same time. However, in some DBMSs such as Microsoft Access, you must run each insert statement one at a time.

Whenever you have a primary key specified in a table, make sure you do not run the same insert statement more than once because the DBMS will generate an error.

Figure 2.3 shows the populated Employee table.

The INSERT statements add five rows of records to the Employees table. The *INSERT* statement is used to insert a single row into a table. The number of insert statements you create determines the number of records in your table.

Each insert statement contains two parts (INSERT INTO, VALUES). The INSERT INTO keywords are used to specify the table name and column names to insert data into. The VALUES keyword is used to specify the values to insert. All of the values are separated by commas and all values that contain text are enclosed in single quotes. Figure

17

*Chapter 2*

2.3 shows the populated Employee table. The DBMS automatically displays the records sorted by the SocialSecNum column.

## Populated Employees Table

| SocialSecNum | Firstname | Lastname | Address | Zipcode | Areacode | PhoneNumber |
|---|---|---|---|---|---|---|
| 226-73-1919 | Jacob | Lincoln | 2609 40th Ave S Honolulu, HI | 96820 | 808 | 423-4111 |
| 249-74-1682 | Jackie | Fields | 2211 Peachtree St N Tampa, FL | 33612 | 813 | 827-2301 |
| 263-73-1442 | Adam | Williams | 1938 32nd Ave S. St. Pete, FL | 33711 | 727 | 321-2234 |
| 266-11-4444 | Sam | Elliot | 1601 Center Loop Tampa, FL | 33612 | 813 | 898-2134 |
| 266-73-1982 | John | Dentins | 2211 22nd Ave N Atlanta, GA | 98718 | 301 | 897-4321 |

**Figure 2.3**

**Inserting a Row Without Specifying Column Names**

# Example 3

It is possible to run INSERT statements without specifying the column names. To accomplish this, you must type the values in the same order as the columns appear in the table. For example, the following record can be inserted into the Employees table without specifying the column names:

INSERT INTO Employees
VALUES ('216-43-1982', 'Peter', 'Lake', '1000 2nd Ave N FL, Tampa', 33611, 301, '897-1217');

**Inserting a Row that is Missing a Column Value**

# Example 4

If you want to insert a record that contains a field (column) with no value, use the NULL keyword.

For example, say you need to enter a new record into the Employees table but you do not have all the information that pertains to the employee. You can use the NULL keyword in place of the missing

*Creating Tables and Working with Data*

values. The following script uses the NULL keyword in the place of a missing area code and phone number:

INSERT INTO Employees
VALUES ('263-13-1002', 'Jacob', 'Tennis', '2000 Florida Ave S FL, Tampa', 33689, NULL, NULL);

**Inserting Values For Specific Columns**

# Example 5

If you want to insert values for only specific columns, specify the column names you want to insert data into. After the name of the table, specify the specific column names. After the VALUES keyword, specify the values for the columns you specified. Look at the following script that demonstrates this:

INSERT INTO Employees (SocialSecNum, FirstName, LastName, Address, Zipcode)
VALUES ('209-74-1602', 'Jackie', 'Fields', '3035 Peachtree St S Tampa, FL', 33612);

There may come a time when you may need to transfer data from one table to another. There are two ways to accomplish this task. The first way involves transferring data to an existing table and the second involves transferring data to a new table. Look at example six and seven.

# Transferring Data from One Table to an Existing Table

In SQL Server, to transfer data to an existing table you must use a new keyword called SELECT. The *SELECT* keyword is primarily used to specify specific column(s) from a table. You will learn more about the SELECT keyword in chapter three. Look at the following example:

*Chapter 2*

# Example 6

Say you want to transfer data from a table named NewEmployees to an existing table named Employees. Look at the following script.

INSERT INTO Employees (SocialSecNum, Firstname, Lastname, Address, Zipcode, Areacode, PhoneNumber)

SELECT SocialSecNum, Firstname, Lastname, Address, Zipcode, Areacode, PhoneNumber
FROM NewEmployees;

In the preceding example, the SELECT keyword is used to tell the DBMS to select the records in the SocialSecNum, Firstname, Lastname, Address, Zipcode, Areacode, and PhoneNumber columns from a table named NewEmployees. The *FROM* keyword is used to tell the DBMS which table to select data from.

The INSERT INTO keywords are used to tell the DBMS to insert the records from the SELECT statement into an existing table named Employees.

To insert records from one table to an existing table both tables do not have to have the same column names, but they do have to have similar datatypes and field sizes specified.

# Transferring Data from One Table to a New Table

# Example 7

Say you want to create a new table on the fly and transfer data from another table to the newly created table. Look at the following script.

```
SELECT *
INTO NewEmployees
FROM Employees;
```

The *asterisk (*)* symbol is often used with the SELECT keyword to tell the DBMS to select every column from a table. In the preceding script, the SELECT and FROM keywords are used in conjunction with an asterisk to tell the DBMS to select all the records from an existing table named Employees.

In most cases, the INTO keyword is used with the INSERT keyword to specify table and column names to insert into a table. In example seven, the INTO keyword is used to create a new table on the fly named NewEmployees.

The new table (NewEmployees) will automatically contain the same datatypes and field sizes that the Employees table contains. After the NewEmployees table is created the DBMS inserts the records from the Employees table into the NewEmployees table.

## Updating Data in a Table

The *UPDATE* statement is used to update one or more rows in a table. Look at example eight, which shows how to update two fields (columns) in the Members table in figure 2.4.

## Example 8

### Members Table

| MemberID | Firstname | Lastname | Address | City | State | Zipcode | Areacode | PhoneNumber |
|---|---|---|---|---|---|---|---|---|
| 1 | Jeffrey | Lindley | 3980 14th Ave S | Atlanta | GA | 98700 | 301 | 451-5451 |
| 2 | Jerry | Lindsey | 4000 3rd Ave S | Tampa | FL | 33600 | 813 | 923-7852 |
| 3 | Gerry | Pitts | 3090 13th St N | Tampa | FL | 33611 | 813 | 286-4821 |
| 4 | Stan | Benson | 1825 8th St N | Santa Fe | NM | 88388 | 505 | 464-1578 |
| 5 | Peter | Gable | 1097 10th Ave S | St. Petersburg | FL | 33754 | 727 | 327-1253 |

## Figure 2.4

*Chapter 2*

Say you want to change Jeffrey Lindley's address and zip code in the Members table. Look at the following script:

```
UPDATE Members
SET
Address = '1290 14th Ave N',
Zipcode = 98722
WHERE MemberID = 1;
```

The preceding script uses the UPDATE keyword to instruct the DBMS to update the Members table. The *SET* keyword is used to assign a new address (1290 14th Ave N) and zip code (98722) to the Members table.

The *WHERE* clause is used to set a condition in a query. It uses a comparison operator (=) to set a condition to update member ID 1. The comparison operator is used to compare each member ID in the Members table to value 1 until it finds a match.

The WHERE clause and the comparison operators are discussed in more detail in chapter 4.

Figure 2.5 shows the Members table with the new address and zip code.

## Updated Members Table

| MemberID | Firstname | Lastname | Address | City | State | Zipcode | Areacode | PhoneNumber |
|---|---|---|---|---|---|---|---|---|
| 1 | Jeffrey | Lindley | 1290 14th Ave N | Atlanta | GA | 98722 | 301 | 451-5451 |
| 2 | Jerry | Lindsey | 4000 3rd Ave S | Tampa | FL | 33600 | 813 | 923-7852 |
| 3 | Gerry | Pitts | 3090 13th St N | Tampa | FL | 33611 | 813 | 286-4821 |
| 4 | Stan | Benson | 1825 8th St N | Santa Fe | NM | 88388 | 505 | 464-1578 |
| 5 | Peter | Gable | 1097 10th Ave S | St. Petersburg | FL | 33754 | 727 | 327-1253 |

## Figure 2.5

## Deleting Data in a Table

The *DELETE* statement is used to remove records (rows) from a table. It enables you to delete single, multiple, or every row from a table. Look at example nine.

## Example 9

Say you want to delete the record that contains member ID 5. Look at the following script.

DELETE FROM Members
WHERE MemberID = 5;

The DELETE keyword is used to instruct the DBMS to delete a record from the Members table. The WHERE clause uses a comparison operator (=) to set a condition to delete member ID 5. The comparison operator is used to compare each member ID in the Members table to value 5 until it finds a match.

Refer to chapter four for more on the WHERE clause and operators used with the WHERE clause.

 The delete statement cannot be used to delete tables or individual cells of data.

## Conclusion

In this chapter, you learned how to create and populate a table using the CREATE TABLE and INSERT keywords. You also learned how to use the SELECT keyword to transfer data to new and existing tables and how to update and delete records in a table.

*Chapter 2*

# Test Your Knowledge of the Chapter

# Quiz 2

1. What symbol is used to select every column in a table?
2. True or False: The delete statement can be used to delete tables.
3. True or False: A field is equivalent to a column.
4. True or False: *NOT NULL* indicates that a field can be left blank when entering data into a table.
5. True or False: The SELECT INTO keywords are used to transfer data to an existing table.

# Assignment 2

Create and populate a table with three records.

# Chapter 3

# Selecting and Retrieving Data

## Introduction

In this chapter, you will learn how to use the SELECT statement to retrieve specific records from a table.

Take a look at the important terms.

## Important Terms:

**AS**: Keyword used to assign an alternate name to a column.

**Concatenation**: A term used to merge values or columns together.

**SELECT**: Keyword used to specify specific column(s) from a table.

**SELECT Statement**: A statement used to retrieve specific column(s) from a table.

**DISTINCT**: Keyword used to display unique values in that column.

**FROM**: Keyword used to specify specific table(s) in a database.

**UNION**: Keyword used to combine two queries to eliminate duplicate records.

**UNION ALL**: Keyword used to combine two queries to show all duplicate records.

*Chapter 3*

 The examples in this chapter were created using Microsoft SQL Server.

# The SELECT Statement

In chapter 2, the SELECT keyword was discussed briefly. The SELECT statement is used to retrieve specific column(s) from a table. Look at the syntax for the SELECT statement.

## SELECT Statement Syntax

SELECT ColumnName, ColumnName, ColumnName
FROM TableName;

The SELECT and FROM keywords are used most often to query a database. They are used to tell the DBMS to look for specific columns and tables stored in a database. The *SELECT* keyword is used to specify specific column(s) from a table and the *FROM* keyword is used to specify specific table(s) in a database.

When you specify columns or tables you must separate each column or table name with a comma and a space.

Look at example one, which shows how to create a SELECT statement to retrieve multiple columns from a table.

 SQL syntax may differ slightly from one DBMS to another.

 SQL statements are not case sensitive. This means the DBMS accepts SQL statements typed in uppercase and lowercase characters.

*Selecting and Retrieving Data*

# Creating a SELECT Statement to Retrieve Multiple Columns

## Employees Table

| SocialSecNum | Firstname | Lastname | Address | Zipcode | Areacode | PhoneNumber |
|---|---|---|---|---|---|---|
| 109-83-4765 | Shaun | Rivers | 1548 6th Ave S Atlanta, GA | 98718 | 301 | 894-1973 |
| 123-88-1982 | Debra | Fields | 1934 16th Ave N Atlanta, GA | 98718 | 301 | 897-3245 |
| 211-73-1112 | Tom | Jetson | 1311 2nd Ave E Atlanta, GA | 98718 | 301 | 897-9877 |
| 226-73-1919 | Jacob | Lincoln | 2609 40th Ave S Honolulu, HI | 96820 | 808 | 423-4111 |
| 249-74-1682 | Jackie | Fields | 2211 Peachtree St N Tampa, FL | 33612 | 813 | 827-2301 |
| 263-73-1442 | Adam | Williams | 1938 32nd Ave S. St. Pete, FL | 33711 | 727 | 321-2234 |
| 266-11-4444 | Sam | Elliot | 1601 Center Loop Tampa, FL | 33612 | 813 | 898-2134 |
| 266-73-1982 | John | Dentins | 2211 22nd Ave N Atlanta, GA | 98718 | 301 | 897-4321 |
| 980-22-1982 | Shawn | Lewis | 1601 4th Ave W Atlanta, GA | 98718 | 301 | 894-0987 |
| 982-24-3490 | Yolanda | Brown | 1544 16th Ave W Atlanta, GA | 98718 | 301 | 892-1234 |

**Figure 3.1**

# Example 1

Say you want to retrieve the first name, last name, and phone number of each employee in the Employees table in figure 3.1. Look at the following script:

SELECT Firstname, Lastname, PhoneNumber
FROM Employees;

The SELECT keyword is used to specify three columns (Firstname, Lastname, and PhoneNumber). The FROM keyword is used to specify the table name (Employees). The closing semicolon is used to tell the DBMS where the query ends.

All DBMSs do not require a closing semicolon. Check your DBMS documentation.

In figure 3.2, three columns (Firstname, Lastname, PhoneNumber) with ten records are retrieved from the Employees table in figure 3.1.

27

*Chapter 3*

## Results (Output)

| Firstname | Lastname | PhoneNumber |
|---|---|---|
| Shaun | Rivers | 894-1973 |
| Debra | Fields | 897-3245 |
| Tom | Jetson | 897-9877 |
| Jacob | Lincoln | 423-4111 |
| Jackie | Fields | 827-2301 |
| Adam | Williams | 321-2234 |
| Sam | Elliot | 898-2134 |
| John | Dentins | 897-4321 |
| Shawn | Lewis | 894-0987 |
| Yolanda | Brown | 892-1234 |

**Figure 3.2**

If you want to display the columns in a different order, type the columns in the SELECT statement in the order you want to display them.

## Creating a SELECT Statement to Retrieve Every Column

## Example 2

Say you want to retrieve every column from a table. Look at the following script:

SELECT *
FROM Employees;

The preceding script uses an asterisk (*) in the SELECT statement. You were first introduced to the asterisk symbol in chapter two. The asterisk (*) is often used to tell the DBMS to select every column in a table. Figure 3.3 shows the output from the preceding script.

## Results (Output)

| SocialSecNum | Firstname | Lastname | Address | Zipcode | Areacode | PhoneNumber |
|---|---|---|---|---|---|---|
| 109-83-4765 | Shaun | Rivers | 1548 6th Ave S Atlanta, GA | 98718 | 301 | 894-1973 |
| 123-88-1982 | Debra | Fields | 1934 16th Ave N Atlanta, GA | 98718 | 301 | 897-3245 |
| 211-73-1112 | Tom | Jetson | 1311 2nd Ave E Atlanta, GA | 98718 | 301 | 897-9877 |
| 226-73-1919 | Jacob | Lincoln | 2609 40th Ave S Honolulu, HI | 96820 | 808 | 423-4111 |
| 249-74-1682 | Jackie | Fields | 2211 Peachtree St N Tampa, FL | 33612 | 813 | 827-2301 |
| 263-73-1442 | Adam | Williams | 1938 32nd Ave S. St. Pete, FL | 33711 | 727 | 321-2234 |
| 266-11-4444 | Sam | Elliot | 1601 Center Loop Tampa, FL | 33612 | 813 | 898-2134 |
| 266-73-1982 | John | Dentins | 2211 22nd Ave N Atlanta, GA | 98718 | 301 | 897-4321 |
| 980-22-1982 | Shawn | Lewis | 1601 4th Ave W Atlanta, GA | 98718 | 301 | 894-0987 |
| 982-24-3490 | Yolanda | Brown | 1544 16th Ave W Atlanta, GA | 98718 | 301 | 892-1234 |

Figure 3.3

# Using DISTINCT to display Unique Values in a Column

# Example 3

If you have a column that contains duplicate values, you can use the *DISTINCT* keyword to display the unique values in that column. Look at the following example for further explanation:

SELECT DISTINCT Areacode
FROM Employees;

The preceding script displays unique area codes from the Employees table. It uses the DISTINCT keyword before the Areacode column to eliminate duplicate values in the Areacode column. Figure 3.4 shows the output.

## Results (Output)

| Areacode |
|---|
| 301 |
| 727 |
| 808 |
| 813 |

**Figure 3.4**

## Using the AS Keyword to Create an Alternate Name for a Column

## Example 4

### Employees Table

| SocialSecNum | Firstname | Lastname | Address | Zipcode | Areacode | PhoneNumber |
|---|---|---|---|---|---|---|
| 109-83-4765 | Shaun | Rivers | 1548 6th Ave S Atlanta, GA | 98718 | 301 | 894-1973 |
| 123-88-1982 | Debra | Fields | 1934 16th Ave N Atlanta, GA | 98718 | 301 | 897-3245 |
| 211-73-1112 | Tom | Jetson | 1311 2nd Ave E Atlanta, GA | 98718 | 301 | 897-9877 |
| 226-73-1919 | Jacob | Lincoln | 2609 40th Ave S Honolulu, HI | 96820 | 808 | 423-4111 |
| 249-74-1682 | Jackie | Fields | 2211 Peachtree St N Tampa, FL | 33612 | 813 | 827-2301 |
| 263-73-1442 | Adam | Williams | 1938 32nd Ave S. St. Pete, FL | 33711 | 727 | 321-2234 |
| 266-11-4444 | Sam | Elliot | 1601 Center Loop Tampa, FL | 33612 | 813 | 898-2134 |
| 266-73-1982 | John | Dentins | 2211 22nd Ave N Atlanta, GA | 98718 | 301 | 897-4321 |
| 980-22-1982 | Shawn | Lewis | 1601 4th Ave W Atlanta, GA | 98718 | 301 | 894-0987 |
| 982-24-3490 | Yolanda | Brown | 1544 16th Ave W Atlanta, GA | 98718 | 301 | 892-1234 |

**Figure 3.5**

When you retrieve columns from a table you can substitute different names for columns. Look at the following script that creates a substitute name for the Address column from the Employees table in figure 3.5.

SELECT Firstname, Lastname, Address AS ContactAddress
FROM Employees;

*Selecting and Retrieving Data*

The preceding script uses the AS keyword to specify an alternate name (ContactAddress) for the Address column. The *AS* keyword is used to assign an alternate name to a column. In figure 3.6 the Address column is renamed to ContactAddress.

## Results (Output)

| Firstname | Lastname | ContactAddress |
|---|---|---|
| Shaun | Rivers | 1548 6th Ave S Atlanta, GA |
| Debra | Fields | 1934 16th Ave N Atlanta, GA |
| Tom | Jetson | 1311 2nd Ave E Atlanta, GA |
| Jacob | Lincoln | 2609 40th Ave S Honolulu, HI |
| Jackie | Fields | 2211 Peachtree St N Tampa, FL |
| Adam | Williams | 1938 32nd Ave S. St. Pete, FL |
| Sam | Elliot | 1601 Center Loop Tampa, FL |
| John | Dentins | 2211 22nd Ave N Atlanta, GA |
| Shawn | Lewis | 1601 4th Ave W Atlanta, GA |
| Yolanda | Brown | 1544 16th Ave W Atlanta, GA |

**Figure 3.6**

The AS keyword does not physically change column names in the table. It is used to enable you to display results under an alternate name.

Some DBMSs do not recognize the AS keyword. In most cases, you can simply separate the column name and alternate column name with one space.

*Chapter 3*

# Merging Columns

# Example 5

In the SELECT statement, you can merge columns together under a newly defined alternate column name. Merging columns is a form of *concatenation*. In Microsoft SQL Server, concatenation requires the use of the following symbol: (+). Look at the following script that illustrates concatenation.

SELECT Firstname + Lastname AS EmployeesName
FROM Employees;

In the preceding script, the Firstname and Lastname columns are merged together using the concatenation symbol. The AS keyword is used to display the merged column as EmployeesName.

Figure 3.7 shows one column named EmployeesName. The EmployeesName column shows the first and last name of each employee from the Employees table.

NOTE | Microsoft Access uses the ampersand (&) symbol to perform concatenation. Oracle uses the double pipe (||) symbol to perform concatenation. MySQL uses a concatenation function (CONCAT ()) to perform concatenation.

## Results (Output)

| EmployeesName | |
|---|---|
| Shaun | Rivers |
| Debra | Fields |
| Tom | Jetson |
| Jacob | Lincoln |
| Jackie | Fields |
| Adam | Williams |
| Sam | Elliot |
| John | Dentins |
| Shawn | Lewis |
| Yolanda | Brown |

**Figure 3.7**

# Uniting Queries to Compare Records in Two Separate Tables

You can use the UNION and UNION ALL keywords to combine two queries to compare results of one table to another. The *UNION* keyword is used to combine two queries to eliminate duplicate records. The *UNION ALL* keywords are used to combine two queries to show all duplicate records. Look at example 6 which shows an example using the UNION keyword.

## UNION

# Example 6

Figures 3.8 and 3.9 shows two tables (Committee1, Committee2). Each table represents a committee that employees at a company belong to. Some employees do not belong to a committee, some belong to only one committee, and some belong to both committees.

## Committee1 Table

| SocialSecNum | Firstname | Lastname | Address | Zipcode | Areacode | PhoneNumber |
|---|---|---|---|---|---|---|
| 123-88-1982 | Debra | Fields | 1934 16th Ave N Atlanta, GA | 98718 | 301 | 897-3245 |
| 211-73-1112 | Tom | Jetson | 1311 2nd Ave E Atlanta, GA | 98718 | 301 | 897-9877 |
| 226-73-1919 | Jacob | Lincoln | 2609 40th Ave S Honolulu, HI | 96820 | 808 | 423-4111 |
| 249-74-1682 | Jackie | Fields | 2211 Peachtree St N Tampa, FL | 33612 | 813 | 827-2301 |
| 263-73-1442 | Adam | Williams | 1938 32nd Ave S. St. Pete, FL | 33711 | 727 | 321-2234 |

**Figure 3.8**

## Committee2 Table

| SocialSecNum | Firstname | Lastname | Address | Zipcode | Areacode | PhoneNumber |
|---|---|---|---|---|---|---|
| 226-73-1919 | Jacob | Lincoln | 2609 40th Ave S Honolulu, HI | 96820 | 808 | 423-4111 |
| 249-74-1682 | Jackie | Fields | 2211 Peachtree St N Tampa, FL | 33612 | 813 | 827-2301 |
| 263-73-1442 | Adam | Williams | 1938 32nd Ave S. St. Pete, FL | 33711 | 727 | 321-2234 |
| 266-11-4444 | Sam | Elliot | 1601 Center Loop Tampa, FL | 33612 | 813 | 898-2134 |
| 266-73-1982 | John | Dentins | 2211 22nd Ave N Atlanta, GA | 98718 | 301 | 897-4321 |

**Figure 3.9**

To find out which employees belong to at least one committee without displaying duplicate names, type the following script:

SELECT Firstname, Lastname
FROM Committee1

UNION

SELECT Firstname, Lastname
FROM Committee2

The preceding script shows two SELECT statements. The first SELECT statement retrieves the first and last names from the Committee1 table. The second SELECT statement retrieves the first and last names from the Committee2 table. The UNION keyword is used to eliminate duplicate names.

*Selecting and Retrieving Data*

Figure 3.10 shows the first and last names of every employee that belongs to at least one committee with no duplicate names.

## Results (Output)

| Firstname | Lastname |
|---|---|
| John | Dentins |
| Sam | Elliot |
| Tom | Jetson |
| Jackie | Fields |
| Jacob | Lincoln |
| Adam | Williams |
| Debra | Fields |

**Figure 3.10**

# UNION ALL

## Example 7

To display employees that belong to a committee including employees that belong to both committees, type the following script:

SELECT Firstname, Lastname
FROM Committee1

UNION ALL

SELECT Firstname, Lastname
FROM Committee2

This example is much like the last example except the duplicate records are included. The UNION ALL keywords enable the DBMS to display every record specified, including duplicates.

Chapter 3

In figure 3.11 Adam Williams, Jackie Fields and Jacob Lincoln are displayed twice. All three employees belong to committee1 and committee2.

## Results (Output)

| Firstname | Lastname |
|---|---|
| Debra | Fields |
| Tom | Jetson |
| Jacob | Lincoln |
| Jackie | Fields |
| Adam | Williams |
| Jacob | Lincoln |
| Jackie | Fields |
| Adam | Williams |
| Sam | Elliot |
| John | Dentins |

## Figure 3.11

NOTE: Some DBMSs support the INTESECT keyword. This keyword enables you to retrieve only the employees that belong to both committees.

# Conclusion

In this chapter you learned how to use the SELECT statement to retrieve records from a table. You learned how to display unique values in a column, create an alternate name for a column, merge two columns, and how to combine queries to compare records in two separate tables.

# Test Your Knowledge of the Chapter

# Quiz 3

1. What keyword is used to create an alternate name for a column?
2. What symbol is used to perform concatenation?
3. True or False: UNION ALL is used to combine two queries to show all duplicates records.
4. True or False: Some DBMSs use the (||) symbol in place of the (+) symbol to perform concatenation.
5. True or False: The FROM keyword is used to specify specific column(s) in a table.

# Assignment 3

Use the Employees table from this chapter to create a query that retrieves the SocialSecNum, Firstname, and Lastname columns. Merge the Firstname and Lastname columns and create an alternate column name for the merged columns.

# Chapter 4

# Filter Retrieved Data

## Introduction

In this chapter, you will learn how to use the WHERE clause to filter retrieved records.

Look at the important terms for this chapter.

## Important Terms:

**WHERE Clause**: Used to filter retrieved records.

**Operators**: Used in the WHERE clause to set conditions on data.

**Expression**: Any data type that returns a value.

The examples in this chapter were created using Microsoft SQL Server.

## The WHERE Clause

The *WHERE* clause is part of the SELECT statement. It is used to filter retrieved records. You can filter retrieved data by setting specific conditions on the data. Look at the syntax for the WHERE clause.

### WHERE Clause Syntax

*Filter Retrieved Data*

WHERE [Condition];

The WHERE clause uses operators to help filter retrieved records. Operators are used in the WHERE clause to help you to set conditions on data. There are four types of operators used in the WHERE clause: comparison, character, logical and shorthand.

## WHERE Clause Operators

### Comparison Operators

The comparison operators are used to compare expressions. An *expression* is any data type that returns a value. Look at figure 4.1.

### Comparison Operators

| !< | Not Less Than | = | Equality |
|---|---|---|---|
| > | Greater Than | <> | Non-Equality |
| >= | Greater Than or Equal To | != | Non-Equality |
| !> | Not Greater Than | < | Less Than |
| IS NULL | No Value | <= | Less Than or Equal To |

**Figure 4.1**

### Character Operators

The character operators are used to perform wildcard-character searches, which use special characters to match parts of a value. Look at the explanations below.

**LIKE**: Used with either the percent (%) or asterisk (*) symbol to match parts of a value.

**Underscore (_)**: Used to match single characters.

## Logical Operators

The logical operators are used to separate two or more conditions. Look at the explanations below.

**AND**: Requires both expressions on either side of the AND operator be true in order for data to be returned.

**NOT**: Used to match any condition opposite of the one defined.

**OR**: Requires at least one expression on either side of the OR operator be true in order for data to be returned.

## Shorthand Operators

These additional operators provide shorthand methods for writing SQL statements. Look at the explanations below.

**BETWEEN**: Provides a shorter method for checking for a range of values.

**IN**: Provides a shorter method for specifying a range of conditions.

# Using a Comparison Operator (=) to Match a Condition

# Example 1

*Filter Retrieved Data*

## Employees Table

| SocialSecNum | Firstname | Lastname | Address | Zipcode | Areacode | PhoneNumber |
|---|---|---|---|---|---|---|
| 109-83-4765 | Shaun | Rivers | 1548 6th Ave S Atlanta, GA | 98718 | 301 | 894-1973 |
| 123-88-1982 | Debra | Fields | 1934 16th Ave N Atlanta, GA | 98718 | 301 | 897-3245 |
| 211-73-1112 | Tom | Jetson | 1311 2nd Ave E Atlanta, GA | 98718 | 301 | 897-9877 |
| 226-73-1919 | Jacob | Lincoln | 2609 40th Ave S Honolulu, HI | 96820 | 808 | 423-4111 |
| 249-74-1682 | Jackie | Fields | 2211 Peachtree St N Tampa, FL | 33612 | 813 | 827-2301 |
| 263-73-1442 | Adam | Williams | 1938 32nd Ave S. St. Pete, FL | 33711 | 727 | 321-2234 |
| 266-11-4444 | Sam | Elliot | 1601 Center Loop Tampa, FL | 33612 | 813 | 898-2134 |
| 266-73-1982 | John | Dentins | 2211 22nd Ave N Atlanta, GA | 98718 | 301 | 897-4321 |
| 980-22-1982 | Shawn | Lewis | 1601 4th Ave W Atlanta, GA | 98718 | 301 | 894-0987 |
| 982-24-3490 | Yolanda | Brown | 1544 16th Ave W Atlanta, GA | 98718 | 301 | 892-1234 |

## Figure 4.2

Say you want to use the Employees table in figure 4.2 to display employee information on employees that have an 813 area code. You could type the following script:

SELECT Lastname, Address, Zipcode, Areacode, PhoneNumber
FROM Employees
WHERE Areacode = 813;

The preceding script uses a comparison operator (=) in the WHERE clause to compare each area code in the Employees table to the value 813 until it finds a match. Figure 4.3 displays two records that contain an 813 area code.

## Results (Output)

| Lastname | Address | Zipcode | Areacode | PhoneNumber |
|---|---|---|---|---|
| Fields | 2211 Peachtree St N Tampa, FL | 33612 | 813 | 827-2301 |
| Elliot | 1601 Center Loop Tampa, FL | 33612 | 813 | 898-2134 |

## Figure 4.3

*Chapter 4*

# Using a Character Operator (%LIKE) to Match a Condition

# Example 2

The *LIKE* operator is used with the percent (%) symbol to match parts of a value.

Say you want to use the Employees table in figure 4.2 to retrieve the last name, first name and phone number for every employee whose last name begins with the letter L. In this case, you could use the LIKE operator.

Look at the following script:

SELECT Lastname, FirstName, PhoneNumber
FROM Employees
WHERE Lastname LIKE 'L%';

The preceding script uses the LIKE operator in the WHERE clause to tell the DBMS to search for every name in the Lastname column that starts with the letter L. The percent (%) symbol is used to specify the letter L. Since the letter L is placed before the percent sign the DBMS knows to search for the last names that begin with the letter L.

To find values that end with a specific character, simply place the character after the percent symbol. For example, to find records that end with the letter L type; '%L'

The percent sign and the letter L are surrounded by quotes. Whenever you specify characters in the WHERE clause they must be surrounded by quotes.

*Filter Retrieved Data*

Figure 4.4 displays two records (Jacob Lincoln and Shawn Lewis).

## Results (Output)

| Lastname | FirstName | PhoneNumber |
|---|---|---|
| Lincoln | Jacob | 423-4111 |
| Lewis | Shawn | 894-0987 |

## Figure 4.4

In Microsoft Access, you must use an asterisk (*) symbol with the LIKE operator to match parts of a value.

## The %LIKE% Operator

The LIKE operator can also be used to find any occurrence of specified characters. In this case, specified characters must be surrounded by two percent symbols. Look at the following example:

# Example 3

Say you want information on employees who live in Tampa. Since the Employees table in figure 4.2 does not have a separate column for the city, you could use a character operator to locate the cities in the address column. Look at the following script:

SELECT Address, Lastname, Firstname
FROM Employees
WHERE Address LIKE '%Tampa%';

The preceding script uses the LIKE operator and two percent symbols (%) in the WHERE clause to tell the DBMS to search for any occurrence of the word Tampa in the Address column.

Chapter 4

Figure 4.5 displays two records; Jackie Fields and Sam Elliot. Jackie Fields and Sam Elliot both live in Tampa.

## Results (Output)

| Address | Lastname | Firstname |
|---|---|---|
| 2211 Peachtree St N Tampa, FL | Fields | Jackie |
| 1601 Center Loop Tampa, FL | Elliot | Sam |

**Figure 4.5**

## The LIKE _ Operator

The underscore symbol is also used with the LIKE operator. The underscore symbol is used as a placeholder when you specify characters of a specific pattern. Look at example four.

# Example 4

Say you want to display phone numbers that match a certain pattern. For example: 894-_9__. Look at the following script:

SELECT Firstname, Lastname, PhoneNumber
FROM Employees
WHERE PhoneNumber LIKE '894-_9__';

The preceding script uses the underscore operator (_) with the LIKE operator to match phone numbers that begin with 894 and the fifth number is 9. The underscore operator is used as a placeholder.

Figure 4.6 displays two records. Both Shaun Rivers and Shawn Lewis match the criterion set in the WHERE clause.

*Filter Retrieved Data*

## Results (Output)

| Firstname | Lastname | PhoneNumber |
|---|---|---|
| Shaun | Rivers | 894-1973 |
| Shawn | Lewis | 894-0987 |

**Figure 4.6**

# Using a Logical Operator (AND, OR) to Match a Condition

The AND and OR operators are logical operators used to separate two or more conditions. The *AND* operator requires both expressions on either side of the AND operator be true in order for data to be returned. The *OR* operator requires at least one expression on either side of the OR operator be true in order for data to be returned. Look at example five.

# Example 5

## Courses Table

| CourseID | StudentID | Course | StartTime | EndTime | StartDate | EndDate | Teacher | Credit |
|---|---|---|---|---|---|---|---|---|
| A1000 | 2 | Accounting I | 2:00pm | 4:00pm | 2/3/2003 | 5/3/2003 | Mrs. Smith | 3 |
| A1001 | 5 | Accounting II | 1:00pm | 3:00pm | 2/3/2003 | 5/3/2003 | Mrs. Terry | 3 |
| D1000 | 3 | Database Basics | 1:00pm | 3:00pm | 2/3/2003 | 5/3/2003 | Mr. Carter | 3 |
| H1011 | 1 | Human Resource Mgt | 3:00pm | 5:00pm | 2/3/2003 | 5/3/2003 | Mr. Pen | 3 |
| L1001 | 3 | Literature | 2:00pm | 4:00pm | 2/3/2003 | 5/3/2003 | Mrs. Donaldson | 3 |
| M1101 | 1 | Pre Algebra | 3:00pm | 5:00pm | 2/3/2003 | 5/3/2003 | Mr. Stevens | 3 |
| M1102 | 5 | Pre Calculus | 3:00pm | 5:00pm | 2/3/2003 | 5/3/2003 | Mr. Dixon | 3 |
| M1103 | 4 | Statistics | 3:00pm | 5:00pm | 2/3/2003 | 5/3/2003 | Mr. Levin | 3 |
| P2000 | 4 | Physics | 2:00pm | 4:00pm | 2/3/2003 | 5/3/2003 | Mrs. Jones | 3 |
| R1001 | 2 | Reading | 1:00pm | 3:00pm | 2/3/2003 | 5/3/2003 | Ms Jackson | 3 |

**Figure 4.7**

Figure 4.7 shows a table named Courses. The Courses table contains information about courses offered at a local college.

*Chapter 4*

Say you want to use the Courses table to display information on courses taught by either Mrs. Smith or Mr. Stevens. Additionally, you only want information on courses that begin at 3:00pm. Look at the following script:

SELECT CourseID, Course, StartTime, EndTime, StartDate, Teacher
FROM Courses
WHERE (Teacher = 'Mrs. Smith' OR Teacher = 'Mr. Stevens') AND StartTime LIKE '%3%';

In the preceding script, the OR and AND operators are used to set conditions on the data to retrieve. The AND operator is always processed before the OR operator. To process the condition containing the OR operator first, it must be surrounded by parenthesis because operators that are surrounded in parentheses are processed first.

The OR operator is used to tell the DBMS to display information for either Mrs. Smith or Mr. Stevens. Notice how both names (Mrs. Smith, Mr. Stevens) are set equal to the Teacher column in the WHERE clause. Finally, since all courses start on the hour, the LIKE operator is used to find any occurrence of the number 3. Figure 4.8 displays the only course that matches the criterion set in the preceding script.

**Results (Output)**

| CourseID | Course | StartTime | EndTime | StartDate | Teacher |
|---|---|---|---|---|---|
| M1101 | Pre Algebra | 3:00pm | 5:00pm | 2003-02-03 00:00:00.000 | Mr. Stevens |

**Figure 4.8**

# Using a Logical Operator (NOT) to Match a Condition Opposite of the One Defined

The *NOT* operator is used to match any condition opposite of the one defined. Look at the following example using the NOT operator:

*Filter Retrieved Data*

# Example 6

Say you want information on every course, excluding the ones that have course ID's that start with the letter M. You can use the NOT operator along with the LIKE operator to accomplish this. Look at the following script.

SELECT CourseID, Course, StartTime, EndTime, StartDate, Teacher
FROM Courses
WHERE CourseID NOT LIKE 'M%';

In the preceding script, the NOT operator is used to retrieve information on every course except the one defined in the WHERE clause. The LIKE operator is used to specify CourseID's starting with the letter M. Figure 4.9 shows the result of the preceding script.

 In many DBMSs, you can place the NOT operator before or after the column name in the WHERE clause.

## Results (Output)

| CourseID | Course | StartTime | EndTime | StartDate | Teacher |
|---|---|---|---|---|---|
| A1000 | Accounting I | 2:00pm | 4:00pm | 2003-02-03 00:00:00.000 | Mrs. Smith |
| A1001 | Accounting II | 1:00pm | 3:00pm | 2003-02-03 00:00:00.000 | Mrs. Terry |
| D1000 | Database Basics | 1:00pm | 3:00pm | 2003-02-03 00:00:00.000 | Mr. Carter |
| H1011 | Human Resource Mgt | 3:00pm | 5:00pm | 2003-02-03 00:00:00.000 | Mr. Pen |
| L1001 | Literature | 2:00pm | 4:00pm | 2003-02-03 00:00:00.000 | Mrs. Donaldson |
| P2000 | Physics | 2:00pm | 4:00pm | 2003-02-03 00:00:00.000 | Mrs. Jones |
| R1001 | Reading | 1:00pm | 3:00pm | 2003-02-03 00:00:00.000 | Ms Jackson |

Figure 4.9

# Using the IN Operator to Match a Condition

The IN operator provides a shorter method for specifying a range of conditions. Look at example seven.

*Chapter 4*

# Example 7

## Courses Table

| CourseID | StudentID | Course | StartTime | EndTime | StartDate | EndDate | Teacher | Credit |
|---|---|---|---|---|---|---|---|---|
| A1000 | 2 | Accounting I | 2:00pm | 4:00pm | 2/3/2003 | 5/3/2003 | Mrs. Smith | 3 |
| A1001 | 5 | Accounting II | 1:00pm | 3:00pm | 2/3/2003 | 5/3/2003 | Mrs. Terry | 3 |
| D1000 | 3 | Database Basics | 1:00pm | 3:00pm | 2/3/2003 | 5/3/2003 | Mr. Carter | 3 |
| H1011 | 1 | Human Resource Mgt | 3:00pm | 5:00pm | 2/3/2003 | 5/3/2003 | Mr. Pen | 3 |
| L1001 | 3 | Literature | 2:00pm | 4:00pm | 2/3/2003 | 5/3/2003 | Mrs. Donaldson | 3 |
| M1101 | 1 | Pre Algebra | 3:00pm | 5:00pm | 2/3/2003 | 5/3/2003 | Mr. Stevens | 3 |
| M1102 | 5 | Pre Calculus | 3:00pm | 5:00pm | 2/3/2003 | 5/3/2003 | Mr. Dixon | 3 |
| M1103 | 4 | Statistics | 3:00pm | 5:00pm | 2/3/2003 | 5/3/2003 | Mr. Levin | 3 |
| P2000 | 4 | Physics | 2:00pm | 4:00pm | 2/3/2003 | 5/3/2003 | Mrs. Jones | 3 |
| R1001 | 2 | Reading | 1:00pm | 3:00pm | 2/3/2003 | 5/3/2003 | Ms Jackson | 3 |

**Figure 4.10**

Suppose you want to retrieve course information for three specific course ID's (M1101, M1102, M1103). Look at the following script.

SELECT CourseID, Course, StartTime, EndTime, StartDate, Teacher
FROM Courses
WHERE CourseID IN ('M1101', 'M1102', 'M1103');

In the preceding script, the IN operator is used to specify three course ID's (M1101, M1102, M1103) in the WHERE clause. Each course ID is surrounded in single quotes since each course ID includes a character.

NOTE

The following script is equivalent to the script in example seven:

SELECT CourseID, Course, StartTime, EndTime, StartDate, Teacher
FROM Courses
WHERE CourseID = 'M1101' OR CourseID = 'M1102' OR CourseID = 'M1103';

Figure 4.11 displays course information for the three course ID's (M1101, M1102, M1103) specified.

## Results (Output)

| CourseID | Course | StartTime | EndTime | StartDate | Teacher |
|---|---|---|---|---|---|
| M1101 | Pre Algebra | 3:00pm | 5:00pm | 2003-02-03 00:00:00.000 | Mr. Stevens |
| M1102 | Pre Calculus | 3:00pm | 5:00pm | 2003-02-03 00:00:00.000 | Mr. Dixon |
| M1103 | Statistics | 3:00pm | 5:00pm | 2003-02-03 00:00:00.000 | Mr. Levin |

**Figure 4.11**

# Using the BETWEEN Operator to Match a Condition

The BETWEEN operator provides a shorter method for checking for a range of values. Look at the following example:

# Example 8

## Employees Table

| SocialSecNum | Firstname | Lastname | Address | Zipcode | Areacode | PhoneNumber |
|---|---|---|---|---|---|---|
| 109-83-4765 | Shaun | Rivers | 1548 6th Ave S Atlanta, GA | 98718 | 301 | 894-1973 |
| 123-88-1982 | Debra | Fields | 1934 16th Ave N Atlanta, GA | 98718 | 301 | 897-3245 |
| 211-73-1112 | Tom | Jetson | 1311 2nd Ave E Atlanta, GA | 98718 | 301 | 897-9877 |
| 226-73-1919 | Jacob | Lincoln | 2609 40th Ave S Honolulu, HI | 96820 | 808 | 423-4111 |
| 249-74-1682 | Jackie | Fields | 2211 Peachtree St N Tampa, FL | 33612 | 813 | 827-2301 |
| 263-73-1442 | Adam | Williams | 1938 32nd Ave S. St. Pete, FL | 33711 | 727 | 321-2234 |
| 266-11-4444 | Sam | Elliot | 1601 Center Loop Tampa, FL | 33612 | 813 | 898-2134 |
| 266-73-1982 | John | Dentins | 2211 22nd Ave N Atlanta, GA | 98718 | 301 | 897-4321 |
| 980-22-1982 | Shawn | Lewis | 1601 4th Ave W Atlanta, GA | 98718 | 301 | 894-0987 |
| 982-24-3490 | Yolanda | Brown | 1544 16th Ave W Atlanta, GA | 98718 | 301 | 892-1234 |

**Figure 4.12**

*Chapter 4*

Say you want to use the Employees table in figure 4.12 to retrieve information on employees who live in or between zip code areas 33612 and 98800. Look at the following script:

SELECT Firstname, Lastname, PhoneNumber
FROM Employees
WHERE Zipcode BETWEEN 33612 AND 98800;

The preceding script retrieves the first name, last name, and phone number of every employee that lives in or between 33612 and 98800 zip code areas. The BETWEEN operator is used to specify a range of numbers (33612 – 98800). The AND operator is used to specify the two numbers the zip code must fall between.

> The following script is equivalent to the script in example eight:
>
> SELECT Firstname, Lastname, PhoneNumber
> FROM Employees
> WHERE Zipcode >= 33612 AND Zipcode <= 98800;

Figure 4.13 produces three columns with ten records as a result of the query.

## Results (Output)

| Firstname | Lastname | PhoneNumber |
|---|---|---|
| Shaun | Rivers | 894-1973 |
| Debra | Fields | 897-3245 |
| Tom | Jetson | 897-9877 |
| Jacob | Lincoln | 423-4111 |
| Jackie | Fields | 827-2301 |
| Adam | Williams | 321-2234 |
| Sam | Elliot | 898-2134 |
| John | Dentins | 897-4321 |
| Shawn | Lewis | 894-0987 |
| Yolanda | Brown | 892-1234 |

## Figure 4.13

*Filter Retrieved Data*

# Conclusion

In this chapter, you learned how to use the comparison, character, logical and shorthand operators in the WHERE clause to filter retrieved records.

# Test Your Knowledge of the Chapter

# Quiz 4

1. Which character operator is used with the percent symbol to match parts of a value?
2. Which type of operator is used to perform wildcard-character searches?
3. True or False: The WHERE clause is used to combine two queries to show all duplicates records.
4. True or False: Operators are used in the WHERE clause to set conditions on data.
5. Which type of operator is used to separate two or more conditions in a WHERE clause?

# Assignment 4

Use the Courses table in figure 4.7 to create a query that shows the courses for students with the following student ID's: 1, 2 and 3

# Chapter 5

# Creating Calculated Fields

## Introduction

In this chapter, you will learn how to incorporate calculated fields into your queries.

Read the important terms before you begin.

## Important Terms:

**Arithmetic operators**: Operators used to perform mathematical calculations.

**Calculated fields**: Used to reformat retrieved data.

**Concatenation**: A term used to merge values or columns together.

**Functions**: Used to reformat report or calculate data.

The examples in this chapter were created using Microsoft SQL Server.

## Calculated Fields

*Calculated fields* are used to reformat data stored in a table. They enable you to manipulate retrieved data. Calculated fields in Structured Query Language (SQL) are created using any of the following: concatenation, arithmetic operators, or functions.

*Creating Calculated Fields*

# Concatenation

In chapter 3, you learned about concatenation. *Concatenation* is a term used to merge values or columns together. Look at the following example that demonstrates concatenation.

## Example 1

## Employees Table

| SocialSecNum | Firstname | Lastname | Address | Zipcode | Areacode | PhoneNumber |
|---|---|---|---|---|---|---|
| 109-83-4765 | Shaun | Rivers | 1548 6th Ave S Atlanta, GA | 98718 | 301 | 894-1973 |
| 123-88-1982 | Debra | Fields | 1934 16th Ave N Atlanta, GA | 98718 | 301 | 897-3245 |
| 211-73-1112 | Tom | Jetson | 1311 2nd Ave E Atlanta, GA | 98718 | 301 | 897-9877 |
| 226-73-1919 | Jacob | Lincoln | 2609 40th Ave S Honolulu, HI | 96820 | 808 | 423-4111 |
| 249-74-1682 | Jackie | Fields | 2211 Peachtree St N Tampa, FL | 33612 | 813 | 827-2301 |
| 263-73-1442 | Adam | Williams | 1938 32nd Ave S. St. Pete, FL | 33711 | 727 | 321-2234 |
| 266-11-4444 | Sam | Elliot | 1601 Center Loop Tampa, FL | 33612 | 813 | 898-2134 |
| 266-73-1982 | John | Dentins | 2211 22nd Ave N Atlanta, GA | 98718 | 301 | 897-4321 |
| 980-22-1982 | Shawn | Lewis | 1601 4th Ave W Atlanta, GA | 98718 | 301 | 894-0987 |
| 982-24-3490 | Yolanda | Brown | 1544 16th Ave W Atlanta, GA | 98718 | 301 | 892-1234 |

## Figure 5.1

Say you want to use the Employees table in figure 5.1 to merge two columns to additional text you typed. Look at the following script:

SELECT Lastname AS Employees, 'lives at '+ Address + Zipcode AS Address
FROM Employees;

The preceding script uses the AS keyword to display the Lastname column as an alternate name (Employees). Next, the Address and Zipcode columns are merged to the additional text surrounded by single quotes ('lives at ') and displayed under an alternate name (Address).

Figure 5.2 displays the two columns (Employees, Address) with ten records.

Chapter 5

## Results (Output)

| Employees | Address | |
|---|---|---|
| Rivers | lives at 1549 6th Ave S Atlanta, GA | 98718 |
| Fields | lives at 1934 16th Ave N Atlanta, GA | 98718 |
| Jetson | lives at 1311 2nd Ave E Atlanta, GA | 98718 |
| Lincoln | lives at 2609 40th Ave S Honolulu, HI | 96820 |
| Fields | lives at 2211 Peachtree St N Tampa, FL | 33612 |
| Williams | lives at 1938 32nd Ave S. St. Pete, FL | 33711 |
| Elliot | lives at 1601 Center Loop Tampa, FL | 33612 |
| Dentins | lives at 2211 22nd Ave N Atlanta, GA | 98718 |
| Lewis | lives at 1601 4th Ave W Atlanta, GA | 98718 |
| Brown | lives at 1544 16th Ave W Atlanta, GA | 98718 |

**Figure 5.2**

NOTE: Microsoft Access uses the ampersand (&) symbol to perform concatenation. Oracle uses the double pipe (||) symbol to perform concatenation. MySQL uses a concatenation function (CONCAT ()) to perform concatenation.

## Arithmetic Operators

*Arithmetic operators* are used to perform mathematical calculations. They are also used to create calculated fields. You can use them to perform addition, subtraction, multiplication, division and remainder calculations. They are pretty much self-explanatory. Look at the arithmetic operators:

**Plus (+)**
**Minus (-)**
**Multiply (\*)**
**Divide (/)**
**Modulus (%)**

# Example 2

## Supplies Table

| SupplyID | SupplyName | Price | SalePrice | InStock | OnOrder |
|---|---|---|---|---|---|
| AR100 | Animated Rainbow | 20 | 18 | 10 | 20 |
| CC100 | Crystal Cat | 75 | 67.5 | 60 | 20 |
| CD100 | China Doll | 20 | 18 | 200 | 0 |
| CP100 | China Puppy | 15 | 13.5 | 20 | 40 |
| DB100 | Dancing Bird | 10 | 9 | 10 | 20 |
| FL100 | Friendly Lion | 14 | 12.6 | 0 | 30 |
| GR100 | Glass Rabbit | 50 | 45 | 50 | 20 |
| MT100 | Miniature Train Set | 60 | 54 | 1 | 30 |
| PS100 | Praying Statue | 25 | 22.5 | 3 | 40 |
| WC100 | Wooden Clock | 11 | 9.9 | 100 | 0 |

## Figure 5.3

Figure 5.3 shows a table named Supplies. The Supplies table contains information about supplies that are sold at a company.

Say you want to use the Supplies table in figure 5.3 to create a new sale price column that calculates a 20% percent discount for every item in the Supplies table. Look at the following script:

SELECT SupplyID, SupplyName, Price, (Price *.80) AS TwentyPercentDiscount
FROM Supplies;

In the preceding script, an arithmetic operator (*) is used to perform multiplication. The Price column is multiplied by 80 percent to show a 20 percent discount.

Figure 5.4 displays four columns (SupplyID, SupplyName, Price, and TwentyPercentDiscount). The TwentyPercentDiscount column shows the price of the supply after the 20 percent discount.

Chapter 5

## Results (Output)

| SupplyID | SupplyName | Price | TwentyPercentDiscount |
|---|---|---|---|
| AR100 | Animated Rainbow | 20 | 16.00 |
| CC100 | Crystal Cat | 75 | 60.00 |
| CD100 | China Doll | 20 | 16.00 |
| CP100 | China Puppy | 15 | 12.00 |
| DB100 | Dancing Bird | 10 | 8.00 |
| FL100 | Friendly Lion | 14 | 11.20 |
| GR100 | Glass Rabbit | 50 | 40.00 |
| MT100 | Miniature Train Set | 60 | 48.00 |
| PS100 | Praying Statue | 25 | 20.00 |
| WC100 | Wooden Clock | 11 | 8.80 |

**Figure 5.4**

# Functions

Functions are also used to create calculated fields. *Functions* are used to reformat, report or calculate data.

One thing you should keep in mind before using functions is that functions in Structured Query Language (SQL) are database specific. This means that they vary from one DBMS to another. For the most part, SQL is pretty standard, but when it comes to functions it is best to check your DBMS documentation for descriptions and uses of functions for your DBMS.

Although functions differ from one DBMS to another, each DBMS usually contains a function that falls into one or more of the following groups of functions: aggregate, arithmetic, date and time, and character. Look at the following explanations:

## Aggregate Functions

The aggregate functions return a single value on values stored in a column.

**AVG ()**: Returns the average of a column.

**COUNT ()**: Counts the number of rows in a column.

**MAX ()**: Returns the highest number in a column.

**MIN ()**: Returns the lowest number in a column.

**SUM ()**: Returns the sum of a column.

## Arithmetic Functions

The arithmetic functions manipulate numeric data.

**EXP ()**: Returns a value of e (exponent) raised to the power of a given number.

**SQRT ()**: Returns the square root of a number.

**ABS ()**: Returns the absolute value of a number.

## Date and Time Functions

The date and time functions manipulate values based on the time and date.

**GETDATE ()**: Returns the current date and time on a computer.

**DATENAME (MONTH, GETDATE ())**: Returns characters that represent a specified datepart of a specified date.

**DATEPART (DATEPART, DATE)**: Returns integers that represent a specified datepart of a specified date.

## Character Functions

The character functions manipulate characters.

**LTRIM ()**: Removes extra spaces from the left of a value.

*Chapter 5*

**REPLACE ()**: Replaces specified data.

**RTRIM ()**: Removes extra spaces from the right of a value.

**Table 5.1.** Aggregate, Arithmetic, Date/Time and Character functions

## Aggregate Functions

Aggregate functions return a single value on values stored in a column. Refer to Table 5.1 for more on aggregate functions. Look at example three.

# Example 3

## Supplies Table

| SupplyID | SupplyName | Price | SalePrice | InStock | OnOrder |
|---|---|---|---|---|---|
| AR100 | Animated Rainbow | 20 | 18 | 10 | 20 |
| CC100 | Crystall Cat | 75 | 67.5 | 60 | 20 |
| CD100 | China Doll | 20 | 18 | 200 | 0 |
| CP100 | China Puppy | 15 | 13.5 | 20 | 40 |
| DB100 | Dancing Bird | 10 | 9 | 10 | 20 |
| FL100 | Friendly Lion | 14 | 12.6 | 0 | 30 |
| GR100 | Glass Rabbit | 50 | 45 | 50 | 20 |
| MT100 | Miniature Train Set | 60 | 54 | 1 | 30 |
| PS100 | Praying Statue | 25 | 22.5 | 3 | 40 |
| WC100 | Wooden Clock | 11 | 9.9 | 100 | 0 |

**Figure 5.5**

Say you want to use the Supplies table in figure 5.5 to find the total amount of supplies in stock, the highest and lowest price charged for an item, the average stock on an item and the total number of items currently for sale. Look at the following script.

SELECT SUM (InStock) AS TotalStock, MAX (Price) AS HighestPrice, MIN (Price) AS LowestPrice, AVG (InStock) AS AverageStockOnItems, COUNT (SupplyName) AS TotalItemsForSale
FROM Supplies;

# Creating Calculated Fields

The preceding script demonstrates the SUM (), MIN (), MAX (), AVG (), and COUNT () aggregate functions. The SUM () function adds every number in the InStock column. The MAX () function locates the highest number in the Price column. The MIN () function locates the lowest number in the Price column. The AVG () function averages the numbers in the InStock column and the COUNT () function counts the number of records in the SupplyID column.

Figure 5.6 displays the result of the newly created columns (TotalStock, HighestPrice, LowestPrice, AverageStockOnItems, and TotalItemsForSale).

## Results (Output)

| TotalStock | HighestPrice | LowestPrice | AverageStockOnItems | TotalItemsForSale |
|---|---|---|---|---|
| 454 | 75.00 | 10.00 | 45 | 10 |

**Figure 5.6**

To count every record in a table use the asterisk (*) symbol in the COUNT (*) function.

## Arithmetic Functions

The arithmetic functions manipulate numeric data. Look at example four.

## Example 4

Chapter 5

# Numbers Table

| Column1 | Column2 | Column3 |
|---|---|---|
| 20 | 4 | 21 |
| 10 | 5 | 20 |
| 30 | 10 | 16 |
| 50 | 2 | 18 |
| 60 | 30 | 12 |
| 70 | 2 | 2 |
| 10 | 39 | 2 |
| 40 | 29 | 19 |
| 80 | 54 | 15 |
| 20 | 66 | 23 |

# Figure 5.7

Figure 5.7 shows a table named Numbers. The Numbers table tracks three columns (Column1, Column2, Column3) of numbers.

Say you want to use the Numbers table in figure 5.7 to create a query using the EXP (), SQRT, and MOD () functions. Look at the following script:

SELECT EXP (Column1) AS Exponent, SQRT (Column2) AS Squareroot, ABS (Column3) AS AbsoluteValue
FROM Numbers;

The preceding script uses the EXP () function to raise e by the values stored in the Column1 column. The SQRT () function squares every value in the Column2 column. The ABS () function returns the absolute value of the values stored in the Column3 column.

Figure 5.8 displays the result of the newly created columns (Exponent, Squareroot, and AbsoluteValue).

*Creating Calculated Fields*

## Results (Output)

| Exponent | Squareroot | AbsoluteValue |
|---|---|---|
| 485165195.40979028 | 2.0 | 21 |
| 22026.465794806718 | 2.2360679774997898 | 20 |
| 10686474581524.463 | 3.1622776601683795 | 16 |
| 5.184705528587072E+21 | 1.4142135623730951 | 18 |
| 1.14200738981568428E+26 | 5.4772255750516612 | 12 |
| 2.5154386709191669E+30 | 1.4142135623730951 | 2 |
| 22026.465794806718 | 6.2449979983983983 | 2 |
| 2.3538526683702E+17 | 5.3851648071345037 | 19 |
| 5.5406223843935098E+34 | 7.3484692283495345 | 15 |
| 485165195.40979028 | 8.1240384046359608 | 23 |

**Figure 5.8**

## Date and Time Functions

The date and time functions manipulate values based on the time and date. Look at example five.

## Example 5

This example uses date and time functions to separately display the current month in text form from a computers system date, the number representing the month from a computers systems date, and the entire system date and time on a computer. Look at the following script.

SELECT DATENAME (month, GETDATE ()) AS CurrentMonth, DATEPART (month, GETDATE ()) AS MonthNumber, GETDATE () AS CurrentDate;

The preceding script nests the GETDATE () function within the DATENAME () function and the DATEPART () function.

*Chapter 5*

The GETDATE () function is used to locate the system time and date. After the system date is located the DATENAME () function displays the month in text form.

The DATEPART () function displays the number of the month from the system date. Finally, the GETDATE () function is used separately to display the entire system time and date.

The CurrentMonth column shows the current month (March) on the computer. The MonthNumber column shows the number (3) of the month of March, and the CurrentDate column shows the current date and time on the computer. The time is 10:54 am. Fifty-one point one zero-zero (51.100) is 51 seconds rounded to the milliseconds.

Figure 5.9 displays three columns (CurrentMonth, MonthNumber, and CurrentDate).

## Results (Output)

| CurrentMonth | MonthNumber | CurrentDate |
|---|---|---|
| March | 3 | 2003-03-08 10:54:51.100 |

**Figure 5.9**

*Creating Calculated Fields*

## Character Functions

The character functions manipulate characters. Look at example six.

# Example 6

### Results From Example 1

| Employees | Address | |
|---|---|---|
| Rivers | lives at 1548 6th Ave S Atlanta, GA | 98718 |
| Fields | lives at 1934 16th Ave N Atlanta, GA | 98718 |
| Jetson | lives at 1311 2nd Ave E Atlanta, GA | 98718 |
| Lincoln | lives at 2609 40th Ave S Honolulu, HI | 96820 |
| Fields | lives at 2211 Peachtree St N Tampa, FL | 33612 |
| Williams | lives at 1938 32nd Ave S. St. Pete, FL | 33711 |
| Elliot | lives at 1601 Center Loop Tampa, FL | 33612 |
| Dentins | lives at 2211 22nd Ave N Atlanta, GA | 98718 |
| Lewis | lives at 1601 4th Ave W Atlanta, GA | 98718 |
| Brown | lives at 1544 16th Ave W Atlanta, GA | 98718 |

### Figure 5.10

Figure 5.10 shows the results from example one. Example one merged two columns to additional typed text ('lives at ').

Say you want to change the results from example one. You want to make a correction to one of the employee names and you want to bring the zip code closer to the address. Look at the following script:

SELECT REPLACE (Lastname, 'Fields', 'Field') AS LastnameCorrection, 'lives at '+ RTRIM (Address) +" "+ Zipcode AS Address
FROM Employees;

The preceding script uses the REPLACE () function. The REPLACE () function enables you to specify three parameters: a column name, the data to change and the data to change it to.

*Chapter 5*

The REPLACE () function changes the name "Fields" in the Lastname column to "Field" The RTRIM () function is used to trim the spaces to the right of the Address column, displaying the address closer to the zip code. Additionally, one space is inserted between the address and the zip code by surrounding a space with two double quotes.

Figure 5.11 shows the name correction (Field) in the LastnameCorrection column and the address and zip code are separated by one space.

## Results (Output)

| LastnameCorrection | Address |
|---|---|
| Rivers | lives at 1548 6th Ave S Atlanta, GA 98718 |
| Field | lives at 1934 16th Ave N Atlanta, GA 98718 |
| Jetson | lives at 1311 2nd Ave E Atlanta, GA 98718 |
| Lincoln | lives at 2609 40th Ave S Honolulu, HI 96820 |
| Field | lives at 2211 Peachtree St N Tampa, FL 33612 |
| Williams | lives at 1938 32nd Ave S. St. Pete, FL 33711 |
| Elliot | lives at 1601 Center Loop Tampa, FL 33612 |
| Dentins | lives at 2211 22nd Ave N Atlanta, GA 98718 |
| Lewis | lives at 1601 4th Ave W Atlanta, GA 98718 |
| Brown | lives at 1544 16th Ave W Atlanta, GA 98718 |

**Figure 5.11**

# Conclusion

In this chapter, you learned how to incorporate calculated fields in queries. You learned how to create concatenated fields, use arithmetic operators and use functions in SQL.

# Test Your Knowledge of the Chapter

# Quiz 5

1. Which arithmetic operator is used to perform multiplication?
2. What function is used to display the system time and date?
3. True or False: The SUM () function counts the number of rows in a column.
4. True or False: The COUNT () function returns the sum of a column.
5. What function replaces specified data?

# Assignment 5

Use the Numbers table in figure 5.7 to create a query that sums Column 1, averages Column 2, and counts Column 3. Display the results using alternate column names.

# Chapter 6

# Using Additional Clauses in Structured Query Language (SQL)

## Introduction

In this chapter, you will learn how to use the ORDER BY, GROUP BY and HAVING clauses.

Read the important terms for this chapter before you begin.

## Important Terms:

**ASC keyword**: Used to sort a column in ascending order.

**Clause**: Used in the SELECT statement to assist in the selection and manipulation of data.

**DESC keyword**: Used to sort a column in descending order.

**GROUP BY Clause**: Used to sort groups of data calculated from aggregate functions.

**HAVING Clause**: Used to set conditions on groups of data that are calculated from aggregate functions.

**ORDER BY Clause**: Used to sort specified columns in ascending or descending order.

 The examples in this chapter were created using Microsoft Access.

*Using Additional Clauses in Structured Query Language (SQL)*

# Clauses

Now that you are familiar with the WHERE clause, you are ready to learn how to use the ORDER BY, GROUP BY and HAVING clauses.

# Using the ORDER BY Clause

## Sorting in Ascending Order

To sort retrieved records in a particular order use the ORDER BY clause. The *ORDER BY* clause sorts columns in ascending or descending order. It enables you to specify specific columns to sort and what direction to sort by (ascending or descending). Look at the following example using the ORDER BY clause:

## Example 1

## Employees Table

| SocialSecNum | Firstname | Lastname | Address | Zipcode | Areacode | PhoneNumber |
|---|---|---|---|---|---|---|
| 109-83-4765 | Shaun | Rivers | 1548 6th Ave S Atlanta, GA | 98718 | 301 | 894-1973 |
| 123-88-1982 | Debra | Fields | 1934 16th Ave N Atlanta, GA | 98718 | 301 | 897-3245 |
| 211-73-1112 | Tom | Jetson | 1311 2nd Ave E Atlanta, GA | 98718 | 301 | 897-9877 |
| 226-73-1919 | Jacob | Lincoln | 2609 40th Ave S Honolulu, HI | 96820 | 808 | 423-4111 |
| 249-74-1682 | Jackie | Fields | 2211 Peachtree St N Tampa, FL | 33612 | 813 | 827-2301 |
| 263-73-1442 | Adam | Williams | 1938 32nd Ave S. St. Pete, FL | 33711 | 727 | 321-2234 |
| 266-11-4444 | Sam | Elliot | 1601 Center Loop Tampa, FL | 33612 | 813 | 898-2134 |
| 266-73-1982 | John | Dentins | 2211 22nd Ave N Atlanta, GA | 98718 | 301 | 897-4321 |
| 980-22-1982 | Shawn | Lewis | 1601 4th Ave W Atlanta, GA | 98718 | 301 | 894-0987 |
| 982-24-3490 | Yolanda | Brown | 1544 16th Ave W Atlanta, GA | 98718 | 301 | 892-1234 |

## Figure 6.1

Say you want to use the Employees table in figure 6.1 to display records sorted by the Lastname column. Look at the following script:

Chapter 6

SELECT Lastname, Firstname, SocialSecNum, Address, Zipcode, Areacode, PhoneNumber
FROM Employees
ORDER BY Lastname;

In the preceding script, the column names after the SELECT keyword are typed in the order in which they will be displayed. In the ORDER BY clause the Lastname column is specified to cause the records to be sorted in ascending order.

Figure 6.2 shows the results. The Lastname column is sorted in ascending order.

## Results (Output)

| Lastname | Firstname | SocialSecNum | Address | Zipcode | Areacode | PhoneNumber |
|---|---|---|---|---|---|---|
| Brown | Yolanda | 982-24-3490 | 1544 16th Ave W Atlanta, GA | 98718 | 301 | 892-1234 |
| Dentins | John | 266-73-1982 | 2211 22nd Ave N Atlanta, GA | 98718 | 301 | 897-4321 |
| Elliot | Sam | 266-11-4444 | 1601 Center Loop Tampa, FL | 33612 | 813 | 898-2134 |
| Fields | Debra | 123-88-1982 | 1934 16th Ave N Atlanta, GA | 98718 | 301 | 897-3245 |
| Fields | Jackie | 249-74-1682 | 2211 Peachtree St N Tampa, FL | 33612 | 813 | 827-2301 |
| Jetson | Tom | 211-73-1112 | 1311 2nd Ave E Atlanta, GA | 98718 | 301 | 897-9877 |
| Lewis | Shawn | 980-22-1982 | 1601 4th Ave W Atlanta, GA | 98718 | 301 | 894-0987 |
| Lincoln | Jacob | 226-73-1919 | 2609 40th Ave S Honolulu, HI | 96820 | 808 | 423-4111 |
| Rivers | Shaun | 109-83-4765 | 1548 6th Ave S Atlanta, GA | 98718 | 301 | 894-1973 |
| Williams | Adam | 263-73-1442 | 1938 32nd Ave S. St. Pete, FL | 33711 | 727 | 321-2234 |

## Figure 6.2

NOTE: You can use any column name in the ORDER BY clause even column names that are not specified after the SELECT keyword.

NOTE: The ORDER BY clause defaults to ascending order unless you specify a different search pattern in the ORDER BY clause. For example, you could specify the ORDER BY clause to sort in descending order. You will learn more about sorting in descending order in example three.

*Using Additional Clauses in Structured Query Language (SQL)*

## Sorting Using the ASC Keyword

# Example 2

Whenever you specify a column name in the ORDER BY clause the DBMS automatically sorts the results in ascending order. Although not common, you can also specifically tell the DBMS to sort in ascending order by using the *ASC* keyword within the ORDER BY clause. Look at the following script:

SELECT Lastname, Firstname, SocialSecNum, Address, Zipcode, Areacode, PhoneNumber
FROM Employees
ORDER BY Lastname ASC;

The preceding script uses the ASC keyword in the ORDER BY clause to tell the DBMS to sort the Lastname column in ascending order.

Figure 6.3 shows the Lastname column sorted in ascending order.

## Results (Output)

| Lastname | Firstname | SocialSecNum | Address | Zipcode | Areacode | PhoneNumber |
|---|---|---|---|---|---|---|
| Brown | Yolanda | 982-24-3490 | 1544 16th Ave W Atlanta, GA | 98718 | 301 | 892-1234 |
| Dentins | John | 266-73-1982 | 2211 22nd Ave N Atlanta, GA | 98718 | 301 | 897-4321 |
| Elliot | Sam | 266-11-4344 | 1601 Center Loop Tampa, FL | 33612 | 813 | 898-2134 |
| Fields | Debra | 123-88-1982 | 1934 16th Ave N Atlanta, GA | 98718 | 301 | 897-3245 |
| Fields | Jackie | 249-74-1682 | 2211 Peachtree St N Tampa, FL | 33612 | 813 | 827-2301 |
| Jetson | Tom | 211-73-1112 | 1311 2nd Ave E Atlanta, GA | 98718 | 301 | 897-9877 |
| Lewis | Shawn | 980-22-1982 | 1601 4th Ave W Atlanta, GA | 98718 | 301 | 894-0987 |
| Lincoln | Jacob | 226-73-1919 | 2609 40th Ave S Honolulu, HI | 96820 | 808 | 423-4111 |
| Rivers | Shaun | 109-83-4765 | 1548 6th Ave S Atlanta, GA | 98718 | 301 | 894-1973 |
| Williams | Adam | 263-73-1442 | 1938 32nd Ave S. St. Pete, FL | 33711 | 727 | 321-2234 |

**Figure 6.3**

## Sorting in Descending Order

# Example 3

*Chapter 6*

Say you want to run the previous query again, but instead of sorting the records in ascending order you want to sort in descending order. Type the following script:

SELECT Lastname, Firstname, SocialSecNum, Address, Zipcode, Areacode, PhoneNumber
FROM Employees
ORDER BY Lastname DESC;

The preceding script is the same as the last query except the *DESC* keyword is used in the ORDER BY clause to sort the Lastname column in descending order.

Figure 6.4 shows the Lastname column sorted in descending order.

## Results (Output)

| Lastname | Firstname | SocialSecNum | Address | Zipcode | Areacode | PhoneNumber |
|---|---|---|---|---|---|---|
| Williams | Adam | 263-73-1442 | 1938 32nd Ave S. St. Pete, FL | 33711 | 727 | 321-2234 |
| Rivers | Shaun | 109-83-4765 | 1548 6th Ave S Atlanta, GA | 98718 | 301 | 894-1973 |
| Lincoln | Jacob | 226-73-1919 | 2609 40th Ave S Honolulu, HI | 96820 | 808 | 423-4111 |
| Lewis | Shawn | 980-22-1982 | 1601 4th Ave W Atlanta, GA | 98718 | 301 | 894-0987 |
| Jetson | Tom | 211-73-1112 | 1311 2nd Ave E Atlanta, GA | 98718 | 301 | 897-9877 |
| Fields | Debra | 123-88-1982 | 1934 16th Ave N Atlanta, GA | 98718 | 301 | 897-3245 |
| Fields | Jackie | 249-74-1682 | 2211 Peachtree St N Tampa, FL | 33612 | 813 | 827-2301 |
| Elliot | Sam | 266-11-4444 | 1601 Center Loop Tampa, FL | 33612 | 813 | 898-2134 |
| Dentins | John | 266-73-1982 | 2211 22nd Ave N Atlanta, GA | 98718 | 301 | 897-4321 |
| Brown | Yolanda | 982-24-3490 | 1544 16th Ave W Atlanta, GA | 98718 | 301 | 892-1234 |

**Figure 6.4**

**Sorting Multiple Columns**

# Example 4

*Using Additional Clauses in Structured Query Language (SQL)*

## Employees Table

| SocialSecNum | Firstname | Lastname | Address | Zipcode | Areacode | PhoneNumber |
|---|---|---|---|---|---|---|
| 109-83-4765 | Shaun | Rivers | 1548 6th Ave S Atlanta, GA | 98718 | 301 | 894-1973 |
| 123-88-1982 | Debra | Fields | 1934 16th Ave N Atlanta, GA | 98718 | 301 | 897-3245 |
| 211-73-1112 | Tom | Jetson | 1311 2nd Ave E Atlanta, GA | 98718 | 301 | 897-9877 |
| 226-73-1919 | Jacob | Lincoln | 2609 40th Ave S Honolulu, HI | 96820 | 808 | 423-4111 |
| 249-74-1682 | Jackie | Fields | 2211 Peachtree St N Tampa, FL | 33612 | 813 | 827-2301 |
| 263-73-1442 | Adam | Williams | 1938 32nd Ave S. St. Pete, FL | 33711 | 727 | 321-2234 |
| 266-11-4444 | Sam | Elliot | 1601 Center Loop Tampa, FL | 33612 | 813 | 898-2134 |
| 266-73-1982 | John | Dentins | 2211 22nd Ave N Atlanta, GA | 98718 | 301 | 897-4321 |
| 980-22-1982 | Shawn | Lewis | 1601 4th Ave W Atlanta, GA | 98718 | 301 | 894-0987 |
| 982-24-3490 | Yolanda | Brown | 1544 16th Ave W Atlanta, GA | 98718 | 301 | 892-1234 |

## Figure 6.5

This example sorts two columns (Lastname, Firstname) in the Employees table in figure 6.5. Since one column contains duplicate values the other column is sorted within each duplicate value. Look at the following script:

SELECT Lastname, Firstname, SocialSecNum, Address, Zipcode, Areacode, PhoneNumber
FROM Employees
ORDER BY Lastname, Firstname;

The preceding script sorts the Lastname column and the Firstname column. First, the Lastname column is sorted in ascending order. Next, because the Lastname column contains duplicate values (Fields), the Firstname column is sorted within each duplicate value in the Lastname column.

Figure 6.6 shows the Firstname column sorted within the Lastname column.

*Chapter 6*

## Results (Output)

| Lastname | Firstname | SocialSecNum | Address | Zipcode | Areacode | PhoneNumber |
|---|---|---|---|---|---|---|
| Brown | Yolanda | 982-24-3490 | 1544 16th Ave W Atlanta, GA | 98718 | 301 | 892-1234 |
| Dentins | John | 266-73-1982 | 2211 22nd Ave N Atlanta, GA | 98718 | 301 | 897-4321 |
| Elliot | Sam | 266-11-4444 | 1601 Center Loop Tampa, FL | 33612 | 813 | 898-2134 |
| Fields | Debra | 123-88-1982 | 1934 16th Ave N Atlanta, GA | 98718 | 301 | 897-3245 |
| Fields | Jackie | 249-74-1682 | 2211 Peachtree St N Tampa, FL | 33612 | 813 | 827-2301 |
| Jetson | Tom | 211-73-1112 | 1311 2nd Ave E Atlanta, GA | 98718 | 301 | 897-9877 |
| Lewis | Shawn | 980-22-1982 | 1601 4th Ave W Atlanta, GA | 98718 | 301 | 894-0987 |
| Lincoln | Jacob | 226-73-1919 | 2609 40th Ave S Honolulu, HI | 96820 | 808 | 423-4111 |
| Rivers | Shaun | 109-83-4765 | 1548 6th Ave S Atlanta, GA | 98718 | 301 | 894-1973 |
| Williams | Adam | 263-73-1442 | 1938 32nd Ave S. St. Pete, FL | 33711 | 727 | 321-2234 |

**Figure 6.6**

> **NOTE** If you sort two columns and the first column contains no duplicate values, the DBMS will only sort the first column.

### Sorting Using Numbers

## Example 5

You can also sort columns by specifying the position of columns in a table or by specifying the position in which columns are typed after the SELECT keyword. Look at the following script which sorts columns based on their position in the Employees table in figure 6.5:

SELECT *
FROM Employees
ORDER BY 3, 2;

The preceding script uses the asterisk (*) after the SELECT keyword to display every column in the Employees table. The numbers in the ORDER BY clause specify the third (Lastname) and second (Firstname) columns in the Employees table. This specification in the ORDER BY clause causes the DBMS to sort the third column first. If

*Using Additional Clauses in Structured Query Language (SQL)*

the third column contains duplicate values, the second column will sort within each duplicate value in the third column.

Figure 6.7 shows the Lastname column sorted in ascending order and the Firstname column sorted within duplicate values in the Lastname column.

## Results (Output)

| SocialSecNum | Firstname | Lastname | Address | Zipcode | Areacode | PhoneNumber |
|---|---|---|---|---|---|---|
| 982-24-3490 | Yolanda | Brown | 1544 16th Ave W Atlanta, GA | 98718 | 301 | 892-1234 |
| 266-73-1982 | John | Dentins | 2211 22nd Ave N Atlanta, GA | 98718 | 301 | 897-4321 |
| 266-11-4444 | Sam | Elliot | 1601 Center Loop Tampa, FL | 33612 | 813 | 898-2134 |
| 123-88-1982 | Debra | Fields | 1934 16th Ave N Atlanta, GA | 98718 | 301 | 897-3245 |
| 249-74-1682 | Jackie | Fields | 2211 Peachtree St N Tampa, FL | 33612 | 813 | 827-2301 |
| 211-73-1112 | Tom | Jetson | 1311 2nd Ave E Atlanta, GA | 98718 | 301 | 897-9877 |
| 980-22-1982 | Shawn | Lewis | 1601 4th Ave W Atlanta, GA | 98718 | 301 | 894-0987 |
| 226-73-1919 | Jacob | Lincoln | 2609 40th Ave S Honolulu, HI | 96820 | 808 | 423-4111 |
| 109-83-4765 | Shaun | Rivers | 1548 6th Ave S Atlanta, GA | 98718 | 301 | 894-1973 |
| 263-73-1442 | Adam | Williams | 1938 32nd Ave S. St. Pete, FL | 33711 | 727 | 321-2234 |

**Figure 6.7**

**NOTE** Although the SocialSecNum column is the first column in the Employees table, the results are sorted by the Lastname and Firstname columns.

# Example 6

When you specify columns after the SELECT keyword, you can also use numbers to identify what columns to sort among the columns specified. To accomplish this, type the number that identifies the position of the column in the SELECT statement. Look at the following script:

SELECT Lastname, Firstname, SocialSecNum, Address, Zipcode, Areacode, PhoneNumber
FROM Employees
ORDER BY 1, 2;

Chapter 6

In the preceding script, the number one identifies the Lastname column and the number two identifies the Firstname column.

Figure 6.8 shows the Firstname column sorted within the Lastname column.

**Results (Output)**

| Lastname | Firstname | SocialSecNum | Address | Zipcode | Areacode | PhoneNumber |
|---|---|---|---|---|---|---|
| Brown | Yolanda | 982-24-3490 | 1544 16th Ave W Atlanta, GA | 98718 | 301 | 892-1234 |
| Dentins | John | 266-73-1982 | 2211 22nd Ave N Atlanta, GA | 98718 | 301 | 897-4321 |
| Elliot | Sam | 266-11-4444 | 1601 Center Loop Tampa, FL | 33612 | 813 | 898-2134 |
| Fields | Debra | 123-88-1982 | 1934 16th Ave N Atlanta, GA | 98718 | 301 | 897-3245 |
| Fields | Jackie | 249-74-1682 | 2211 Peachtree St N Tampa, FL | 33612 | 813 | 827-2301 |
| Jetson | Tom | 211-73-1112 | 1311 2nd Ave E Atlanta, GA | 98718 | 301 | 897-9877 |
| Lewis | Shawn | 980-22-1982 | 1601 4th Ave W Atlanta, GA | 98718 | 301 | 894-0987 |
| Lincoln | Jacob | 226-73-1919 | 2609 40th Ave S Honolulu, HI | 96820 | 808 | 423-4111 |
| Rivers | Shaun | 109-83-4765 | 1548 6th Ave S Atlanta, GA | 98718 | 301 | 894-1973 |
| Williams | Adam | 263-73-1442 | 1938 32nd Ave S. St. Pete, FL | 33711 | 727 | 321-2234 |

**Figure 6.8**

# Using the GROUP BY Clause

The *GROUP BY* clause works much like the ORDER BY clause except the GROUP BY clause is used to sort groups of data calculated from aggregate functions. You cannot use the ORDER BY clause in a query that uses an aggregate function. Aggregate functions were discussed in chapter five. Remember, a*ggregate* functions return a single value on values stored in a column.

When you use the GROUP BY clause, every column name you specify to display after the SELECT keyword must also be present in the GROUP BY clause. This only applies to column names physically stored in a table. This does not include alternate column names created using the AS keyword. Look at the following example:

# Example 7

*Using Additional Clauses in Structured Query Language (SQL)*

## Sales Table

| SalesID | SupplyID | CustomerID | DateSold |
|---|---|---|---|
| 1 | AR100 | 2 | 2/3/2003 |
| 2 | WC100 | 8 | 2/5/2003 |
| 3 | AR100 | 7 | 2/6/2003 |
| 4 | FL100 | 1 | 2/8/2003 |
| 5 | MT100 | 3 | 2/8/2003 |
| 6 | GR100 | 4 | 2/10/2003 |
| 7 | WC100 | 5 | 2/22/2003 |
| 8 | PS100 | 9 | 2/20/2003 |
| 9 | CD100 | 6 | 2/18/2003 |
| 10 | CP100 | 10 | 2/17/2003 |
| 11 | CP100 | 10 | 2/17/2003 |
| 12 | CP100 | 5 | 2/17/2003 |
| 13 | CC100 | 4 | 2/17/2003 |
| 14 | GR100 | 3 | 2/8/2003 |
| 15 | MT100 | 2 | 2/17/2003 |
| 16 | WC100 | 1 | 2/8/2003 |
| 17 | CP100 | 3 | 2/8/2003 |

## Figure 6.9

Figure 6.9 shows a table named Sales. The Sales table tracks the sales at a business.

Say you want to use the Sales table in figure 6.9 to display the total amount of items purchased by each customer. You also want to sort the results by the CustomerID column. Look at the following script:

SELECT CustomerID, COUNT (SupplyID) AS TotalPurchases
FROM Sales
GROUP BY CustomerID;

The preceding script uses the COUNT () function to count the Supply ID's for each Customer ID. The result from the function is displayed as TotalPurchases. The GROUP BY clause groups the CustomerID's in ascending order.

Figure 6.10 shows the CustomerID and TotalPurchases column. The results are sorted by the CustomerID column.

**Results (Output)**

| CustomerID | TotalPurchases |
|---|---|
| 1 | 2 |
| 2 | 2 |
| 3 | 3 |
| 4 | 2 |
| 5 | 2 |
| 6 | 1 |
| 7 | 1 |
| 8 | 1 |
| 9 | 1 |
| 10 | 2 |

**Figure 6.10**

# Using the GROUP BY and WHERE Clause

The WHERE clause can also be used with the GROUP BY clause. When you use the GROUP BY clause with the WHERE clause, the GROUP BY clause must be placed after the WHERE clause. Look at example eight.

# Example 8

Say you want to retrieve the most recent order date for three customers. You also want to sort the output by the CustomerID column. Look at the following script:

SELECT CustomerID, MAX (DateSold) AS LastOrderDate
FROM Sales
WHERE CustomerID IN (4, 2, 3)

GROUP BY CustomerID;

In the preceding script, the CustomerID column is specified after the SELECT keyword. The MAX () function is used to retrieve the most recent date from the DateSold column for each CustomerID. The condition in the WHERE clause causes the DBMS to only retrieve information for CustomerID four, two and three. Finally, the GROUP BY clause sorts the results by the CustomerID column.

Figure 6.11 shows two columns (CustomerID, LastOrderDate) with three records (2, 3, and 4). The CustomerID column is sorted in ascending order.

### Results (Output)

| CustomerID | LastOrderDate |
|---|---|
| 2 | 2/17/2003 |
| 3 | 2/8/2003 |
| 4 | 2/17/2003 |

**Figure 6.11**

## Using the HAVING Clause

The *HAVING* clause is used to set conditions on groups of data calculated from aggregate functions. Although you can use the WHERE clause with the GROUP BY clause, the HAVING clause is usually used in place of the WHERE clause since both clauses deal with groups. Whenever you use the HAVING clause, you must also use the GROUP BY clause. Look at example nine.

**NOTE** You can use the WHERE clause with the GROUP BY clause, but you cannot use an aggregate function within a WHERE clause.

*Chapter 6*

> When you use the HAVING clause it must be placed after the GROUP BY clause.

# Example 9

## Sales Table

| SalesID | SupplyID | CustomerID | DateSold |
|---|---|---|---|
| 1 | AR100 | 2 | 2/3/2003 |
| 2 | WC100 | 8 | 2/5/2003 |
| 3 | AR100 | 7 | 2/6/2003 |
| 4 | FL100 | 1 | 2/8/2003 |
| 5 | MT100 | 3 | 2/8/2003 |
| 6 | GR100 | 4 | 2/10/2003 |
| 7 | WC100 | 5 | 2/22/2003 |
| 8 | PS100 | 9 | 2/20/2003 |
| 9 | CD100 | 6 | 2/18/2003 |
| 10 | CP100 | 10 | 2/17/2003 |
| 11 | CP100 | 10 | 2/17/2003 |
| 12 | CP100 | 5 | 2/17/2003 |
| 13 | CC100 | 4 | 2/17/2003 |
| 14 | GR100 | 3 | 2/8/2003 |
| 15 | MT100 | 2 | 2/17/2003 |
| 16 | WC100 | 1 | 2/8/2003 |
| 17 | CP100 | 3 | 2/8/2003 |

## Figure 6.12

Say you want to retrieve the total number of items purchased by customers who purchased two or more items. Look at the following script:

SELECT CustomerID, COUNT (*) AS Purchases
FROM Sales
GROUP BY CustomerID
HAVING Count (*) >=2;

In the preceding script, the CustomerID column is specified after the SELECT keyword. The COUNT () function is used to count and retrieve all the records for each CustomerID. The HAVING clause sets a condition to display only the CustomerID's with a count of two or greater. The GROUP BY clause groups the data by the CustomerID column.

Figure 6.13 displays two columns (CustomerID, Purchases) with six records.

### Results (Output)

| CustomerID | Purchases |
|---|---|
| 1 | 2 |
| 2 | 2 |
| 3 | 3 |
| 4 | 2 |
| 5 | 2 |
| 10 | 2 |

**Figure 6.13**

# Conclusion

In this chapter, you learned how to sort and filter data using the ORDER BY, GROUP BY and HAVING clauses. You learned how to sort in ascending and descending order, sort multiple columns, sort using numbers, and how to combine the GROUP BY clause with the WHERE clause.

# Test Your Knowledge of the Chapter

# Quiz 6

1. What keyword is used to sort a column in descending order?
2. What clause is used to sort groups of data calculated from aggregate functions?

*Chapter 6*

3. True or False: You must use the ASC keyword to sort columns in ascending order.
4. True or False: Aggregate functions can be used in the WHERE clause.
5. True or False: The ORDER BY clause is used to sort specified columns in ascending or descending order.

# Assignment 6

Use the Sales table in figure 6.12 to create a query that locates the first order date for each customer. Group the results by the CustomerID column.

# Chapter 7

# Creating Table Joins

## Introduction

So far you have learned how to retrieve records from a single table. In this chapter, you will learn how to retrieve information from two or more tables simultaneously using table joins.

Look at the important terms for this chapter.

## Important Terms:

**Cartesian product**: Causes each row in the first table to be multiplied by the total number of rows in the second table.

**Inner Join**: Matches values of a column in one table to matching values in another table.

**Joins**: Enables you to use a single SELECT statement to query two or more tables at the same time.

**Keys**: Used to uniquely identify a row or record within a table.

**LEFT OUTER JOIN**: Used to select all the records of data to the left of the LEFT OUTER JOIN keywords in the FROM clause.

**Natural Join**: Used to specify only unique columns from multiple tables.

**ON**: Used to specify a condition.

*Chapter 7*

**Outer Join**: Used to display every record from a table, even if a record is not included in the joined table.

**Qualification**: A term used to show which table a column belongs to. It involves combining a table name with a column name, so there is no question as to which table the column name refers to.

**RIGHT OUTER JOIN**: Used to select all the records of data to the right of the RIGHT OUTER JOIN keywords in the FROM clause.

**Self-join**: Enables you to join a table to itself.

 The examples in this chapter were created using Microsoft Access.

## Table Joins

Table *joins* enable you to use a single SELECT statement to query two or more tables at the same time. Each table you query must contain a key value. *Keys* (primary key, foreign key) uniquely identify a record within a table and they are used to point to specific records in a table. To refresh your memory on keys, refer to chapter one.

## Qualification

Since some tables have matching column names due to key relationships, it is important for you to make it clear as to which table to retrieve a column from. This is called *qualification*. To qualify a table and column name, simply type the table name, a period, and the column name. Look at the following syntax which demonstrates the format to follow to qualify a table and column name:

*Creating Table Joins*

## Qualification Syntax

TableName.ColumnName

There are basically four types of joins used in SQL: Inner Join, Self-Join, Natural join, and Outer join.

 The syntax for joins differs from one DBMS to another. Check your DBMS documentation for changes in syntax.

# Inner Join

An inner join is often referred to as an equi-join. Equi-join refers to equality. An *inner join* is based on equality because it matches values of a column in one table to matching values in another table.

When you create a join, specify the column names you want to retrieve, the tables to retrieve the records from, and the relationships between tables.

If you identify two tables and you do not show the relationship between the two tables, you end up with a Cartesian product. A *Cartesian product* causes each row in the first table to be multiplied by the total number of rows in the second table.

Look at the following example using an inner join:

# Example 1

Chapter 7

## Customers Table

| CustomerID | Firstname | Lastname | Address | City | State | Zipcode | Areacode | PhoneNumber |
|---|---|---|---|---|---|---|---|---|
| 1 | Tom | Evans | 3000 2nd Ave S | Atlanta | GA | 98718 | 301 | 232-9000 |
| 2 | Larry | Genes | 1100 23rd Ave S | Tampa | FL | 33618 | 813 | 982-3455 |
| 3 | Sherry | Jones | 100 Free St S | Tampa | FL | 33618 | 813 | 890-4231 |
| 4 | April | Jones | 2110 10th St S | Santa Fe | NM | 88330 | 505 | 434-1111 |
| 5 | Jerry | Jones | 798 22nd Ave S | St. Pete | FL | 33711 | 727 | 327-3323 |
| 6 | John | Little | 1500 Upside Loop N | St. Pete | FL | 33711 | 727 | 346-1234 |
| 7 | Gerry | Lexingtion | 5642 5th Ave S | Atlanta | GA | 98718 | 301 | 832-8912 |
| 8 | Henry | Denver | 8790 8th St N | Holloman | NM | 88330 | 505 | 423-8900 |
| 9 | Nancy | Kinn | 4000 22nd St S | Atlanta | GA | 98718 | 301 | 879-2345 |
| 10 | Derick | Penns | 2609 15th Ave N | Tampa | FL | 33611 | 813 | 346-1232 |

**Figure 7.1**

## Sales Table

| SalesID | SupplyID | CustomerID | DateSold |
|---|---|---|---|
| 1 | AR100 | 2 | 2/3/2003 |
| 2 | WC100 | 8 | 2/5/2003 |
| 3 | AR100 | 7 | 2/6/2003 |
| 4 | FL100 | 1 | 2/8/2003 |
| 5 | MT100 | 3 | 2/8/2003 |
| 6 | GR100 | 4 | 2/10/2003 |
| 7 | WC100 | 5 | 2/22/2003 |
| 8 | PS100 | 9 | 2/20/2003 |
| 9 | CD100 | 6 | 2/18/2003 |
| 10 | CP100 | 10 | 2/17/2003 |
| 11 | CP100 | 10 | 2/17/2003 |
| 12 | CP100 | 5 | 2/17/2003 |
| 13 | CC100 | 4 | 2/17/2003 |
| 14 | GR100 | 3 | 2/8/2003 |
| 15 | MT100 | 2 | 2/17/2003 |
| 16 | WC100 | 1 | 2/8/2003 |
| 17 | CP100 | 3 | 2/8/2003 |

**Figure 7.2**

Say you want to use the Customers table in figure 7.1 and the Sales table in figure 7.2 to retrieve the customer names and the items each customer purchased. You want to display the Firstname and Lastname columns from the Customers table, and the SupplyID from the Sales

*Creating Table Joins*

table. You also want the results sorted by the Lastname and Firstname columns. Look at the following script:

```
SELECT Lastname, Firstname, SupplyID
FROM Customers, Sales
WHERE Customers.CustomerID = Sales.CustomerID
ORDER BY Lastname, Firstname;
```

In the preceding script, the Lastname, Firstname and SupplyID columns are specified after the SELECT keyword. The FROM clause specifies two table names (Customers, Sales) separated by a comma. The WHERE clause shows the relationship between the Customers and Sales table.

To show the relationship between the Customers and Sales table the column names are qualified. Qualification was discussed in the beginning of this chapter. The Customers and Sales table are related through the CustomerID column because each table contains a CustomerID column. The CustomerID column is the primary key in the Customers table and a foreign key in the Sales table.

Figure 7.3 shows the customer names and the items they purchased. The output is sorted by the Lastname and Firstname columns.

## Results (Output)

| Lastname | Firstname | SupplyID |
|---|---|---|
| Denver | Henry | WC100 |
| Evans | Tom | WC100 |
| Evans | Tom | FL100 |
| Genes | Larry | MT100 |
| Genes | Larry | AR100 |
| Jones | April | CC100 |
| Jones | April | GR100 |
| Jones | Jerry | CP100 |
| Jones | Jerry | WC100 |
| Jones | Sherry | GR100 |
| Jones | Sherry | CP100 |
| Jones | Sherry | MT100 |
| Kinn | Nancy | PS100 |
| Lexingtion | Gerry | AR100 |
| Little | John | CD100 |
| Penns | Derick | CP100 |
| Penns | Derick | CP100 |

**Figure 7.3**

Some DBMSs use slightly different syntax; check your DBMS documentation for changes.

## Self Join

A *Self-join* is used to join a table to itself. The self-join is particularly useful for checking the consistency of data in a table.

Remember, when you create a join; specify the column names you want to retrieve, the tables to retrieve the records from, and the relationships between tables. Look at the following example using a self-join:

# Example 2

## Supplies Table

| SupplyID | SupplyName | Price | SalePrice | InStock | OnOrder |
|---|---|---|---|---|---|
| AR100 | Animated Clouds | $20.00 | $18.00 | 50 | 20 |
| AR100 | Animated Rainbow | $20.00 | $18.00 | 10 | 20 |
| CC100 | Crystal Cat | $75.00 | $67.50 | 60 | 20 |
| CD100 | China Doll | $20.00 | $18.00 | 200 | 0 |
| DB100 | Dancing Bird | $10.00 | $9.00 | 10 | 20 |
| FL100 | Friendly Lion | $14.00 | $12.60 | 0 | 30 |
| GR100 | Glass Rabbit | $50.00 | $45.00 | 50 | 20 |
| MT100 | Miniature Train Set | $60.00 | $54.00 | 1 | 30 |
| PS100 | Praying Statue | $25.00 | $22.50 | 3 | 40 |
| WC100 | Wooden Clock | $11.00 | $9.90 | 100 | 0 |

## Figure 7.4

Figure 7.4 shows a table named Supplies. The Supplies table contains price information for supplies sold at a store.

To demonstrate the self-join, the primary key in the Supplies table was disabled. Additionally, the Supplies table contains two duplicate supply ID's (AR100) with different supply names.

Say you want to create a query that will display duplicate supply ID's with different supply names. Look at the following script:

SELECT s1.SupplyID, s1.SupplyName,
s2.SupplyID, s2.SupplyName
FROM Supplies AS s1, Supplies AS s2
WHERE s1.SupplyID = s2.SupplyID
AND s1.SupplyName <> s2.SupplyName;

*Chapter 7*

In the preceding script, two alternate names or aliases (s1, s2) are created for the Supplies table in the FROM clause.

 To join a table to it self, you must specify table aliases to reflect two separate tables to the DBMS.

After the SELECT keyword the alternate table names are combined with column names (s1.SupplyID, s1.SupplyName, and s2.SupplyID, s2.SupplyName) to qualify each column and table name.

The WHERE clause sets the SupplyID equal (=) to itself and the SupplyName unequal (<>) to itself to find duplicate supply ID's with different supply names.

Figure 7.5 displays the four column names specified after the SELECT keyword in the query. Two records are displayed because there are only two records in the Supplies table that have a matching supply ID and a different supply name.

Since the table is joined to itself it attaches the same two columns to the right of the output. This self-join enables you to easily identify a discrepancy by enabling you to compare column s1.SupplyID to column s2.SupplyID and to compare column s1.SupplyName to column s2.SupplyName.

## Results (Output)

| s1.SupplyID | s1.SupplyName | s2.SupplyID | s2.SupplyName |
|---|---|---|---|
| AR100 | Animated Rainbow | AR100 | Animated Clouds |
| AR100 | Animated Clouds | AR100 | Animated Rainbow |

**Figure 7.5**

## Natural Join

The *Natural join* is used to specify only unique columns from multiple tables. It enables you to eliminate multiple columns from your output. Look at the following example.

# Example 3

### Sales Table

| SalesID | SupplyID | CustomerID | DateSold |
|---|---|---|---|
| 1 | AR100 | 2 | 2/3/2003 |
| 2 | WC100 | 8 | 2/5/2003 |
| 3 | AR100 | 7 | 2/6/2003 |
| 4 | FL100 | 1 | 2/8/2003 |
| 5 | MT100 | 3 | 2/8/2003 |
| 6 | GR100 | 4 | 2/10/2003 |
| 7 | WC100 | 5 | 2/22/2003 |
| 8 | PS100 | 9 | 2/20/2003 |
| 9 | CD100 | 6 | 2/18/2003 |
| 10 | CP100 | 10 | 2/17/2003 |
| 11 | CP100 | 10 | 2/17/2003 |
| 12 | CP100 | 5 | 2/17/2003 |
| 13 | CC100 | 4 | 2/17/2003 |
| 14 | GR100 | 3 | 2/8/2003 |
| 15 | MT100 | 2 | 2/17/2003 |
| 16 | WC100 | 1 | 2/8/2003 |
| 17 | CP100 | 3 | 2/8/2003 |

**Figure 7.6**

Chapter 7

## Customers Table

| CustomerID | Firstname | Lastname | Address | City | State | Zipcode | Areacode | PhoneNumber |
|---|---|---|---|---|---|---|---|---|
| 1 | Tom | Evans | 3000 2nd Ave S | Atlanta | GA | 98718 | 301 | 232-9000 |
| 2 | Larry | Genes | 1100 23rd Ave S | Tampa | FL | 33618 | 813 | 982-3455 |
| 3 | Sherry | Jones | 100 Free St S | Tampa | FL | 33618 | 813 | 890-4231 |
| 4 | April | Jones | 2110 10th St S | Santa Fe | NM | 88330 | 505 | 434-1111 |
| 5 | Jerry | Jones | 798 22nd Ave S | St. Pete | FL | 33711 | 727 | 327-3323 |
| 6 | John | Little | 1500 Upside Loop N | St. Pete | FL | 33711 | 727 | 346-1234 |
| 7 | Gerry | Lexingtion | 5642 5th Ave S | Atlanta | GA | 98718 | 301 | 832-8912 |
| 8 | Henry | Denver | 8790 8th St N | Holloman | NM | 88330 | 505 | 423-8900 |
| 9 | Nancy | Kinn | 4000 22nd St S | Atlanta | GA | 98718 | 301 | 879-2345 |
| 10 | Derick | Penns | 2609 15th Ave N | Tampa | FL | 33611 | 813 | 346-1232 |

Figure 7.7

## Supplies Table

| SupplyID | SupplyName | Price | SalePrice | InStock | OnOrder |
|---|---|---|---|---|---|
| AR100 | Animated Rainbow | $20.00 | $18.00 | 10 | 20 |
| CC100 | Crystal Cat | $75.00 | $67.50 | 60 | 20 |
| CD100 | China Doll | $20.00 | $18.00 | 200 | 0 |
| CP100 | China Puppy | $15.00 | $13.50 | 20 | 40 |
| DB100 | Dancing Bird | $10.00 | $9.00 | 10 | 20 |
| FL100 | Friendly Lion | $14.00 | $12.60 | 0 | 30 |
| GR100 | Glass Rabbit | $50.00 | $45.00 | 50 | 20 |
| MT100 | Miniature Train Set | $60.00 | $54.00 | 1 | 30 |
| PS100 | Praying Statue | $25.00 | $22.50 | 3 | 40 |
| WC100 | Wooden Clock | $11.00 | $9.90 | 100 | 0 |

Figure 7.8

Say you want to retrieve information on customer's who purchased supply ID CP100. You want to retrieve every column from the Sales table in figure 7.6, the Firstname and Lastname columns from the Customers table in figure 7.7, and the SupplyName column from the Supplies table in figure 7.8. Look at the following script.

SELECT S.*, C.Firstname, C.Lastname, SP.SupplyName
FROM Sales AS S, Customers AS C, Supplies AS SP

*Creating Table Joins*

```
WHERE S.CustomerID = C.CustomerID
AND S.SupplyID = SP.SupplyID
AND SP.SupplyID = 'CP100';
```

The preceding script creates three alternate table names (S, C, SP) for the Sales, Customers and Supplies table in the FROM clause. After the SELECT keyword the alternate table names are combined with the column names (Firstname, Lastname, SupplyName) to qualify each column and table name.

The asterisk is used to retrieve every column from the Sales table. The Firstname and Lastname columns are retrieved from the Customers table, and the SupplyName column is retrieved from the Supplies table.

The WHERE clause shows how each column and table specified after the SELECT keyword are related. It additionally sets a condition to specify only the SupplyID's that equal CP100.

Figure 7.9 shows every column from the Sales table, the first and last name columns from the Customers table, and the SupplyName column from the Supplies table.

## Results (Output)

| SalesID | SupplyID | CustomerID | DateSold | Firstname | Lastname | SupplyName |
|---|---|---|---|---|---|---|
| 10 | CP100 | 10 | 2/17/2003 | Derick | Penns | China Puppy |
| 11 | CP100 | 10 | 2/17/2003 | Derick | Penns | China Puppy |
| 12 | CP100 | 5 | 2/17/2003 | Jerry | Jones | China Puppy |
| 17 | CP100 | 3 | 2/8/2003 | Sherry | Jones | China Puppy |

**Figure 7.9**

Chapter 7

# Outer Join

The *Outer join* is used to display every record from a table, even if a record is not included in the joined table. There are two types of outer joins: left outer join and right outer join.

The *LEFT OUTER JOIN* is used to select every record from a table specified to the left of the LEFT OUTER JOIN keywords in the FROM clause. The *RIGHT OUTER JOIN* is used to select every record specified to the right of the RIGHT OUTER JOIN keywords in the FROM clause. Look at the following examples using outer joins:

## LEFT OUTER JOIN

## Example 4

### Customers2 Table

| CustomerID | Firstname | Lastname | Address | City | State | Zipcode | Areacode | PhoneNumber |
|---|---|---|---|---|---|---|---|---|
| 1 | Tom | Evans | 3000 2nd Ave S | Atlanta | GA | 98718 | 301 | 232-9000 |
| 2 | Larry | Genes | 1100 23rd Ave S | Tampa | FL | 33618 | 813 | 982-3455 |
| 3 | Sherry | Jones | 100 Free St S | Tampa | FL | 33618 | 813 | 890-4231 |
| 4 | April | Jones | 2110 10th St S | Santa Fe | NM | 88330 | 505 | 434-1111 |
| 5 | Jerry | Jones | 798 22nd Ave S | St. Pete | FL | 33711 | 727 | 327-3323 |
| 6 | John | Little | 1500 Upside Loop N | St. Pete | FL | 33711 | 727 | 346-1234 |
| 7 | Gerry | Lexingtion | 5642 5th Ave S | Atlanta | GA | 98718 | 301 | 832-8912 |
| 8 | Henry | Denver | 8790 8th St N | Holloman | NM | 88330 | 505 | 423-8900 |
| 9 | Nancy | Kinn | 4000 22nd St S | Atlanta | GA | 98718 | 301 | 879-2345 |
| 10 | Derick | Penns | 2609 15th Ave N | Tampa | FL | 33611 | 813 | 346-1232 |
| 11 | Adam | Williams | 1333 5th St N | Tampa | FL | 33611 | 813 | 326-7777 |
| 12 | Stan | Willows | 1837 30th Ave S | Tampa | FL | 33611 | 813 | 346-1100 |
| 13 | Ricky | Canton | 1009 50th Ave N | Tampa | FL | 33611 | 813 | 346-3223 |
| 14 | Pete | West | 2000 4th Ave N | Tampa | FL | 33611 | 813 | 346-8778 |

Figure 7.10

*Creating Table Joins*

## Sales Table

| SalesID | SupplyID | CustomerID | DateSold |
|---|---|---|---|
| 1 | AR100 | 2 | 2/3/2003 |
| 2 | WC100 | 8 | 2/5/2003 |
| 3 | AR100 | 7 | 2/6/2003 |
| 4 | FL100 | 1 | 2/8/2003 |
| 5 | MT100 | 3 | 2/8/2003 |
| 6 | GR100 | 4 | 2/10/2003 |
| 7 | WC100 | 5 | 2/22/2003 |
| 8 | PS100 | 9 | 2/20/2003 |
| 9 | CD100 | 6 | 2/18/2003 |
| 10 | CP100 | 10 | 2/17/2003 |
| 11 | CP100 | 10 | 2/17/2003 |
| 12 | CP100 | 5 | 2/17/2003 |
| 13 | CC100 | 4 | 2/17/2003 |
| 14 | GR100 | 3 | 2/8/2003 |
| 15 | MT100 | 2 | 2/17/2003 |
| 16 | WC100 | 1 | 2/8/2003 |
| 17 | CP100 | 3 | 2/8/2003 |

## Figure 7.11

Say you want to use figure 7.10 and 7.11 to retrieve customer ID's for customers who made purchases and customer ID's for customers who are on the mailing list but have not made any purchases.

Look at the following script:

SELECT Customers2.CustomerID, Sales.SupplyID
FROM Customers2 LEFT OUTER JOIN Sales
ON Customers2.CustomerID = Sales.CustomerID;

In the preceding script, the CustomerID column and the SupplyID column are qualified after the SELECT keyword. The FROM clause uses the LEFT OUTER JOIN keywords to retrieve all the records in the table (Customers2) specified to the left of the LEFT OUTER JOIN keywords. The ON keyword is used much like the WHERE

*Chapter 7*

clause. It is used to show the relationship between the Customers2 and the Sales table.

 When using the OUTER JOIN keywords, you must use the ON keyword to specify the condition.

Figure 7.12 shows the customer ID for customers who made purchases and for customers that are only on the mailing list. Supply ID's are listed with the appropriate customer ID's.

*Creating Table Joins*

## Results (Output)

| CustomerID | SupplyID |
|---|---|
| 1 | FL100 |
| 1 | WC100 |
| 2 | MT100 |
| 2 | AR100 |
| 3 | GR100 |
| 3 | CP100 |
| 3 | MT100 |
| 4 | CC100 |
| 4 | GR100 |
| 5 | WC100 |
| 5 | CP100 |
| 6 | CD100 |
| 7 | AR100 |
| 8 | WC100 |
| 9 | PS100 |
| 10 | CP100 |
| 10 | CP100 |
| 11 | |
| 12 | |
| 13 | |
| 14 | |

**Figure 7.12**

> The syntax for joins differs from one DBMS to another. Check your DBMS documentation for changes in syntax.

|  **NOTE** | Some DBMSs use the following format to retrieve the same results in the last example:<br><br>SELECT Customers2.CustomerID, Sales.SupplyID<br>FROM Customers2, Sales<br>WHERE Customers2.CustomerID *= Sales.CustomerID;<br><br>The preceding script uses a left outer join operator (*=) in the WHERE clause to return all the records from the table (Customers2) to the left of the left out join operator. |
|---|---|

# RIGHT OUTER JOIN

The *RIGHT OUTER JOIN* is used to select every record from the table specified to the right of the RIGHT OUTER JOIN keywords in the FROM clause. Look at the following example.

# Example 5

## Supplies Table

| SupplyID | SupplyName | Price | SalePrice | InStock | OnOrder |
|---|---|---|---|---|---|
| AR100 | Animated Rainbow | $20.00 | $18.00 | 10 | 20 |
| CC100 | Crystal Cat | $75.00 | $67.50 | 60 | 20 |
| CD100 | China Doll | $20.00 | $18.00 | 200 | 0 |
| CP100 | China Puppy | $15.00 | $13.50 | 20 | 40 |
| DB100 | Dancing Bird | $10.00 | $9.00 | 10 | 20 |
| FL100 | Friendly Lion | $14.00 | $12.60 | 0 | 30 |
| GR100 | Glass Rabbit | $50.00 | $45.00 | 50 | 20 |
| MT100 | Miniature Train Set | $60.00 | $54.00 | 1 | 30 |
| PS100 | Praying Statue | $25.00 | $22.50 | 3 | 40 |
| WC100 | Wooden Clock | $11.00 | $9.90 | 100 | 0 |

**Figure 7.13**

## Sales Table

| SalesID | SupplyID | CustomerID | DateSold |
|---|---|---|---|
| 1 | AR100 | 2 | 2/3/2003 |
| 2 | WC100 | 8 | 2/5/2003 |
| 3 | AR100 | 7 | 2/6/2003 |
| 4 | FL100 | 1 | 2/8/2003 |
| 5 | MT100 | 3 | 2/8/2003 |
| 6 | GR100 | 4 | 2/10/2003 |
| 7 | WC100 | 5 | 2/22/2003 |
| 8 | PS100 | 9 | 2/20/2003 |
| 9 | CD100 | 6 | 2/18/2003 |
| 10 | CP100 | 10 | 2/17/2003 |
| 11 | CP100 | 10 | 2/17/2003 |
| 12 | CP100 | 5 | 2/17/2003 |
| 13 | CC100 | 4 | 2/17/2003 |
| 14 | GR100 | 3 | 2/8/2003 |
| 15 | MT100 | 2 | 2/17/2003 |
| 16 | WC100 | 1 | 2/8/2003 |
| 17 | CP100 | 3 | 2/8/2003 |

## Figure 7.14

Suppose you want to retrieve every customer ID with a corresponding supply ID from the Sales table in figure 7.14. Additionally, you want to retrieve the supply ID's for supplies not purchased yet from the Supplies table in figure 7.13. Look at the following script:

SELECT Sales.CustomerID, Supplies.SupplyID
FROM Sales RIGHT OUTER JOIN Supplies
ON Sales.SupplyID = Supplies.SupplyID;

In the preceding script, the CustomerID column and the SupplyID column are qualified after the SELECT keyword. The FROM clause uses the RIGHT OUTER JOIN keywords to retrieve all the records in the table (Supplies) specified to the right of the RIGHT OUTER JOIN keywords. The ON keyword works like the WHERE clause and is

*Chapter 7*

used to show the relationship between the Sales and the Supplies table.

Figure 7.15 shows the CustomerID column and every Supply ID from the Supplies table. Even the supplies (DB100) that have never been purchased are displayed.

## Results (Output)

| CustomerID | SupplyID |
|---|---|
|  | DB100 |
| 1 | FL100 |
| 1 | WC100 |
| 2 | MT100 |
| 2 | AR100 |
| 3 | GR100 |
| 3 | CP100 |
| 3 | MT100 |
| 4 | GR100 |
| 4 | CC100 |
| 5 | WC100 |
| 5 | CP100 |
| 6 | CD100 |
| 7 | AR100 |
| 8 | WC100 |
| 9 | PS100 |
| 10 | CP100 |
| 10 | CP100 |

**Figure 7.15**

*Creating Table Joins*

>  **NOTE** Some DBMSs display the preceding example as follows:
>
> SELECT Sales.CustomerID, Supplies.SupplyID
> FROM Sales, Supplies
> WHERE Sales.SupplyID =* Supplies.SupplyID;
>
> The preceding script uses a right outer join operator (=*) to return all the records to the right of the operator.

## Conclusion

In this chapter, you learned how to create table joins to retrieve information from two or more tables simultaneously. You learned how to create an inner join, self-join, natural join, left outer join and right outer join.

## Test Your Knowledge of the Chapter

## Quiz 7

1. True or False: Joins enable you to use multiple SELECT statements to query two or more tables at the same time.
2. True or False: An Inner Join matches values of a column in one table to matching values in the same table.
3. Name the two types of outer joins.
4. Which type of join is used to specify only unique columns from multiple tables?
5. Which keyword works like the WHERE clause and is used with the OUTER JOIN keywords?

## Assignment 7

Use the Customers table in figure 7.1 and the Sales table in figure 7.2 to create an inner join displaying the Lastname and SalesID columns.

# Chapter 8

# Creating Subqueries

## Introduction

In this chapter, you will learn how to create and implement subqueries to retrieve records from one or more tables simultaneously. You will also learn how subqueries provide alternative ways to phrase a query.

Read the important terms for this chapter.

## Important Terms:

**ALL**: Used to search for an exact match between values from a query and every value from a subquery.

**ANY**: Used to search for at least one matching value between values from a query and values from a subquery.

**Correlated Subquery**: A subquery that contains a reference to a query outside the subquery.

**EXISTS**: Used to test for the existence or absence of records from a subquery

**IN:** Used to specify a range of conditions.

**Query**: A request or command to the DBMS.

**Subquery**: A query linked to another query permitting values to be passed among queries.

*Creating Subqueries*

**Non-Correlated Subquery**: A subquery that does not contain a reference to a query outside the subquery.

 The examples in this chapter were created using Microsoft Access.

## Subqueries

Subqueries provide you with an additional method to retrieve records from multiple tables. They also provide you with a number of ways to get the same result. A *subquery* is a query linked to another query permitting values to be passed among queries.

Subqueries are usually nested inside other queries or linked to one another using one of the following keywords: IN, EXISTS, ANY or ALL. Subqueries that are linked using keywords process the last subquery first, working upward. Subqueries that are nested inside other queries are processed beginning with the innermost SELECT statement, working outward. As each query is processed, the next query takes the results from the one processed before it. The process continues until every query is processed.

### Correlated and Non-Correlated Subqueries

There are two types of subqueries: correlated and non-correlated. A *correlated* subquery contains a reference to a query outside the subquery. Correlated subqueries execute once for each record a referenced query returns. A *non-correlated* subquery does not contain a reference to a query outside the subquery and therefore only executes once.

When you create subqueries, you must show the relationship between the tables you retrieve data from and every subquery must be surrounded in parentheses.

Chapter 8

 **NOTE** The syntax for subqueries differs from one DBMS to another. Check your DBMS documentation for changes in syntax.

## Using the IN keyword to link Queries

The IN keyword was introduced in chapter four. Recall that it is used to provide a shorter method for specifying a range of conditions. In example one, the IN keyword is used to specify a non-correlated subquery. Look at example one.

## Example 1

### Sales Table

| SalesID | SupplyID | CustomerID | DateSold |
|---|---|---|---|
| 1 | AR100 | 2 | 2/3/2003 |
| 2 | WC100 | 8 | 2/5/2003 |
| 3 | AR100 | 7 | 2/6/2003 |
| 4 | FL100 | 1 | 2/8/2003 |
| 5 | MT100 | 3 | 2/8/2003 |
| 6 | GR100 | 4 | 2/10/2003 |
| 7 | WC100 | 5 | 2/22/2003 |
| 8 | PS100 | 9 | 2/20/2003 |
| 9 | CD100 | 6 | 2/18/2003 |
| 10 | CP100 | 10 | 2/17/2003 |
| 11 | CP100 | 10 | 2/17/2003 |
| 12 | CP100 | 5 | 2/17/2003 |
| 13 | CC100 | 4 | 2/17/2003 |
| 14 | GR100 | 3 | 2/8/2003 |
| 15 | MT100 | 2 | 2/17/2003 |
| 16 | WC100 | 1 | 2/8/2003 |
| 17 | CP100 | 3 | 2/8/2003 |

**Figure 8.1**

*Creating Subqueries*

## Customers Table

| CustomerID | Firstname | Lastname | Address | City | State | Zipcode | Areacode | PhoneNumber |
|---|---|---|---|---|---|---|---|---|
| 1 | Tom | Evans | 3000 2nd Ave S | Atlanta | GA | 98718 | 301 | 232-9000 |
| 2 | Larry | Genes | 1100 23rd Ave S | Tampa | FL | 33618 | 813 | 982-3455 |
| 3 | Sherry | Jones | 100 Free St S | Tampa | FL | 33618 | 813 | 890-4231 |
| 4 | April | Jones | 2110 10th St S | Santa Fe | NM | 88330 | 505 | 434-1111 |
| 5 | Jerry | Jones | 798 22nd Ave S | St. Pete | FL | 33711 | 727 | 327-3323 |
| 6 | John | Little | 1500 Upside Loop N | St. Pete | FL | 33711 | 727 | 346-1234 |
| 7 | Gerry | Lexingtion | 5642 5th Ave S | Atlanta | GA | 98718 | 301 | 832-8912 |
| 8 | Henry | Denver | 8790 8th St N | Holloman | NM | 88330 | 505 | 423-8900 |
| 9 | Nancy | Kinn | 4000 22nd St S | Atlanta | GA | 98718 | 301 | 879-2345 |
| 10 | Derick | Penns | 2609 15th Ave N | Tampa | FL | 33611 | 813 | 346-1232 |

## Figure 8.2

Say you want to use the Sales table in figure 8.1 and the Customers table in figure 8.2 to retrieve the customer ID, first name, and last name of every customer that purchased items with a supply ID that begins with the letter C. Look at the following script:

```
SELECT CustomerID, Firstname, Lastname
FROM Customers
WHERE CustomerID
IN
(SELECT CustomerID
FROM Sales
WHERE SupplyID LIKE 'C*');
```

The preceding script contains a non-correlated subquery linked to another query using the IN keyword. The subquery is processed first and is only executed once. It retrieves customer ID's from the Sales table for Supply ID's that begin with the letter C. Working upward, the next query uses the Customer ID's retrieved from the subquery processed first to retrieve the customer's CustomerID, Firstname, and Lastname columns from the Customers table.

The *IN* keyword is used to specify the subquery. As you can see, the IN keyword is perfect for specifying subqueries. The Customers and

Chapter 8

Sales table are related through the CustomerID column (WHERE CustomerID IN (SELECT CustomerID).

Figure 8.3 shows five records and three columns (CustomerID, Firstname, Lastname) from the Customers table.

## Results (Output)

| CustomerID | Firstname | Lastname |
|---|---|---|
| 3 | Sherry | Jones |
| 4 | April | Jones |
| 5 | Jerry | Jones |
| 6 | John | Little |
| 10 | Derick | Penns |

**Figure 8.3**

## Using the EXISTS keyword to link Queries

The *EXISTS* keyword is used to test for the existence or absence of records from a subquery. It returns TRUE to the DBMS if a subquery returns any records and FALSE if a subquery does not return any results. Look at example two.

# Example 2

Say you want to use the Sales table in figure 8.1 and the Customers table in figure 8.2 to retrieve all the customer information for the customers that purchased Supply ID MT100. Look at the following script:

```
SELECT *
FROM Customers
WHERE EXISTS
(SELECT *
FROM Sales
```

*Creating Subqueries*

WHERE Customers.CustomerID = Sales.CustomerID AND
SupplyID = 'MT100');

The preceding script contains a correlated subquery linked to a query using the EXISTS keyword. The subquery is processed first and is executed once for every record retrieved from the preceding query.

The WHERE clause in the subquery shows the reference to the Customers table in the query (WHERE Customers.CustomerID = Sales.CustomerID). EXISTS returns TRUE for every record that satisfies the condition in the subquery (WHERE Customers.CustomerID = Sales.CustomerID AND SupplyID = 'MT100')).

Figure 8.4 shows two records retrieved from the main query.

## Results (Output)

| CustomerID | Firstname | Lastname | Address | City | State | Zipcode | Areacode | PhoneNumber |
|---|---|---|---|---|---|---|---|---|
| 2 | Larry | Genes | 1100 23rd Ave S | Tampa | FL | 33618 | 813 | 982-3455 |
| 3 | Sherry | Jones | 100 Free St S | Tampa | FL | 33618 | 813 | 890-4231 |

**Figure 8.4**

## Using the ANY keyword to link Queries

The *ANY* keyword is used to search for at least one matching value between values from a query and values from a subquery. It returns TRUE to the DBMS if the comparison operator (>, <, <>, = ANY) is true for any value in the subquery. Look at example three.

Chapter 8

# Example 3

Say you want to use the Customers table in figure 8.2 to retrieve the customer's customer ID, first name, last name and city for any customer that lives in Florida. Look at the following script.

SELECT CustomerID, Firstname, Lastname, City
FROM Customers
WHERE City = ANY
(SELECT City
FROM Customers
WHERE State = 'FL');

The preceding script contains a non-correlated subquery linked to a query using the equality comparison operator (=) and the ANY keyword. The subquery is processed first and is executed once. ANY returns TRUE if values in the main query are equal to any values in the subquery.

| NOTE | The ANY keyword is used much like the IN keyword except IN cannot be used with comparison operators such as =, < and >. |

Figure 8.5 shows five records and four columns (CustomerID, Firstname, Lastname, and City).

## Results (Output)

| CustomerID | Firstname | Lastname | City |
|---|---|---|---|
| 2 | Larry | Genes | Tampa |
| 3 | Sherry | Jones | Tampa |
| 5 | Jerry | Jones | St. Pete |
| 6 | John | Little | St. Pete |
| 10 | Derick | Penns | Tampa |

## Figure 8.5

*Creating Subqueries*

## Using the ALL keyword to link Queries

The *ALL* keyword is used to search for an exact match between values from a query and every value from a subquery. It returns TRUE to the DBMS if the comparison operator (>, <, <>, = ALL) is true for all values in the subquery. Look at example four.

# Example 4

## Customers Table

| CustomerID | Firstname | Lastname | Address | City | State | Zipcode | Areacode | PhoneNumber |
|---|---|---|---|---|---|---|---|---|
| 1 | Tom | Evans | 3000 2nd Ave S | Atlanta | GA | 98718 | 301 | 232-9000 |
| 2 | Larry | Genes | 1100 23rd Ave S | Tampa | FL | 33618 | 813 | 982-3455 |
| 3 | Sherry | Jones | 100 Free St S | Tampa | FL | 33618 | 813 | 890-4231 |
| 4 | April | Jones | 2110 10th St S | Santa Fe | NM | 88330 | 505 | 434-1111 |
| 5 | Jerry | Jones | 798 22nd Ave S | St. Pete | FL | 33711 | 727 | 327-3323 |
| 6 | John | Little | 1500 Upside Loop N | St. Pete | FL | 33711 | 727 | 346-1234 |
| 7 | Gerry | Lexingtion | 5642 5th Ave S | Atlanta | GA | 98718 | 301 | 832-8912 |
| 8 | Henry | Denver | 8790 8th St N | Holloman | NM | 88330 | 505 | 423-8900 |
| 9 | Nancy | Kinn | 4000 22nd St S | Atlanta | GA | 98718 | 301 | 879-2345 |
| 10 | Derick | Penns | 2609 15th Ave N | Tampa | FL | 33611 | 813 | 346-1232 |

**Figure 8.6**

Say you want to use the Customers table in figure 8.6 to retrieve ALL the customers who have a zip code higher than every zip code in Florida. Look at the following script.

SELECT CustomerID, Firstname, Lastname, State, Zipcode
FROM Customers
WHERE Zipcode > ALL
(SELECT Zipcode
FROM Customers
WHERE State = 'FL');

*Chapter 8*

The preceding script uses the ALL keyword to link a query to a non-correlated subquery. The ALL keyword is used in conjunction with the greater than comparison operator (>).

The subquery is processed first and is executed once. ALL returns TRUE if values in the main query are greater than all the values in the subquery.

Figure 8.7 shows five columns (CustomerID, Firstname, Lastname, State, and Zipcode) and five records.

## Results (Output)

| CustomerID | Firstname | Lastname | State | Zipcode |
|---|---|---|---|---|
| 1 | Tom | Evans | GA | 98718 |
| 4 | April | Jones | NM | 88330 |
| 7 | Gerry | Lexingtion | GA | 98718 |
| 8 | Henry | Denver | NM | 88330 |
| 9 | Nancy | Kinn | GA | 98718 |

**Figure 8.7**

# Creating Subqueries to Query Three Tables

# Example 5

## Customers Table

| CustomerID | Firstname | Lastname | Address | City | State | Zipcode | Areacode | PhoneNumber |
|---|---|---|---|---|---|---|---|---|
| 1 | Tom | Evans | 3000 2nd Ave S | Atlanta | GA | 98718 | 301 | 232-9000 |
| 2 | Larry | Genes | 1100 23rd Ave S | Tampa | FL | 33618 | 813 | 982-3455 |
| 3 | Sherry | Jones | 100 Free St S | Tampa | FL | 33618 | 813 | 890-4231 |
| 4 | April | Jones | 2110 10th St S | Santa Fe | NM | 88330 | 505 | 434-1111 |
| 5 | Jerry | Jones | 798 22nd Ave S | St. Pete | FL | 33711 | 727 | 327-3323 |
| 6 | John | Little | 1500 Upside Loop N | St. Pete | FL | 33711 | 727 | 346-1234 |
| 7 | Gerry | Lexingtion | 5642 5th Ave S | Atlanta | GA | 98718 | 301 | 832-8912 |
| 8 | Henry | Denver | 8790 8th St N | Holloman | NM | 88330 | 505 | 423-8900 |
| 9 | Nancy | Kinn | 4000 22nd St S | Atlanta | GA | 98718 | 301 | 879-2345 |
| 10 | Derick | Penns | 2609 15th Ave N | Tampa | FL | 33611 | 813 | 346-1232 |

Figure 8.8

Chapter 8

## Sales Table

| SalesID | SupplyID | CustomerID | DateSold |
|---|---|---|---|
| 1 | AR100 | 2 | 2/3/2003 |
| 2 | WC100 | 8 | 2/5/2003 |
| 3 | AR100 | 7 | 2/6/2003 |
| 4 | FL100 | 1 | 2/8/2003 |
| 5 | MT100 | 3 | 2/8/2003 |
| 6 | GR100 | 4 | 2/10/2003 |
| 7 | WC100 | 5 | 2/22/2003 |
| 8 | PS100 | 9 | 2/20/2003 |
| 9 | CD100 | 6 | 2/18/2003 |
| 10 | CP100 | 10 | 2/17/2003 |
| 11 | CP100 | 10 | 2/17/2003 |
| 12 | CP100 | 5 | 2/17/2003 |
| 13 | CC100 | 4 | 2/17/2003 |
| 14 | GR100 | 3 | 2/8/2003 |
| 15 | MT100 | 2 | 2/17/2003 |
| 16 | WC100 | 1 | 2/8/2003 |
| 17 | CP100 | 3 | 2/8/2003 |

**Figure 8.9**

## Supplies Table

| SupplyID | SupplyName | Price | SalePrice | InStock | OnOrder |
|---|---|---|---|---|---|
| AR100 | Animated Rainbow | $20.00 | $18.00 | 10 | 20 |
| CC100 | Crystal Cat | $75.00 | $67.50 | 60 | 20 |
| CD100 | China Doll | $20.00 | $18.00 | 200 | 0 |
| CP100 | China Puppy | $15.00 | $13.50 | 20 | 40 |
| DB100 | Dancing Bird | $10.00 | $9.00 | 10 | 20 |
| FL100 | Friendly Lion | $14.00 | $12.60 | 0 | 30 |
| GR100 | Glass Rabbit | $50.00 | $45.00 | 50 | 20 |
| MT100 | Miniature Train Set | $60.00 | $54.00 | 1 | 30 |
| PS100 | Praying Statue | $25.00 | $22.50 | 3 | 40 |
| WC100 | Wooden Clock | $11.00 | $9.90 | 100 | 0 |

**Figure 8.10**

*Creating Subqueries*

Say you want to use figures 8.8, 8.9, and 8.10 to retrieve sales information on all the items bought by customers who live in Tampa, Florida. Look at the following script:

SELECT SupplyID, SupplyName, Price, SalePrice
FROM Supplies
WHERE SupplyID

IN

(SELECT SupplyID
FROM Sales
WHERE CustomerID

IN

(SELECT CustomerID
FROM Customers
WHERE City ='Tampa' AND State = 'FL'));

The preceding script uses the IN keyword to link a query to two non-correlated subqueries. The last subquery retrieves the customer ID from the Customers table for every customer that lives in Tampa, Florida.

Moving upward, the next subquery retrieves the SupplyID's from the Sales table that match the Customer ID's retrieved from the query processed before it.

Finally, the last query retrieves the SupplyID, SupplyName, Price, and SalePrice from the Supplies table that match the Supply ID's retrieved from the query processed before it.

Notice the location of the parentheses. This is the format you must use to enclose two subqueries. Each subquery must contain an opening and closing parenthesis.

Chapter 8

Figure 8.11 shows four records and four columns (SupplyID, SupplyName, Price, and SalePrice).

## Results (Output)

| SupplyID | SupplyName | Price | SalePrice |
|---|---|---|---|
| CP100 | China Puppy | $15.00 | $13.50 |
| GR100 | Glass Rabbit | $50.00 | $45.00 |
| MT100 | Miniature Train Set | $60.00 | $54.00 |
| AR100 | Animated Rainbow | $20.00 | $18.00 |

**Figure 8.11**

# Creating Subqueries that Contain Aggregate Functions

Aggregate functions were discussed in chapter five. They return a single value on values stored in a column. There are five aggregate functions: AVG (), COUNT (), MIN (), MAX (), and SUM ().

Subqueries that contain aggregate functions are many times nested within other queries. When subqueries are nested within other queries the innermost SELECT statement, is processed first, working outward.

# Example 6

Say you want to use the Customers and Sales tables in figures 8.8 and 8.9 to retrieve each customer's customer ID and the most recent date each customer purchased an item. Look at the following script:

SELECT CustomerID,
(SELECT MAX (DateSold)
FROM Sales
WHERE Sales.CustomerID = Customers.CustomerID) AS
MostRecentPurchase

FROM Customers
ORDER BY CustomerID;

The preceding script contains a correlated subquery nested inside another query. The WHERE clause in the correlated subquery specifies the reference (Sales.CustomerID = Customers.CustomerID) to the Customers table. The WHERE clause also specifies an alternate column name (MostRecentPurchase) to display the date. The subquery uses the MAX () function to find the most recent date each customer purchased an item. The subquery is executed once for every customer retrieved from the Customers table.

When you nest a subquery, the alias is defined at the end of the subquery and outside of the parentheses.

The outer query retrieves the CustomerID's from the Customers table and sorts the output by the CustomerID column.

In the first line of the script there is a comma after the CustomerID column. The comma tells the DBMS to expect another column name (MostRecentPurchase).

Figure 8.12 displays ten records and two columns (CustomerID, MostRecentPurchase).

Chapter 8

## Results (Output)

| CustomerID | MostRecentPurchase |
|---|---|
| 1 | 2/8/2003 |
| 2 | 2/17/2003 |
| 3 | 2/8/2003 |
| 4 | 2/17/2003 |
| 5 | 2/22/2003 |
| 6 | 2/18/2003 |
| 7 | 2/6/2003 |
| 8 | 2/5/2003 |
| 9 | 2/20/2003 |
| 10 | 2/17/2003 |

**Figure 8.12**

# Example 7

Say you want to use the Sales and Supplies tables in figures 8.9 and 8.10 to display product information and the total amount sold of each product. Look at the following script:

```
SELECT SupplyID, SupplyName, Price, SalePrice,
(SELECT COUNT (*)
FROM Sales
WHERE Sales.SupplyID = Supplies.SupplyID) AS TotalItemsSold
FROM Supplies
ORDER BY SupplyID;
```

The preceding script contains a correlated subquery nested inside another query. The WHERE clause in the correlated subquery shows the reference (Sales.SupplyID = Supplies.SupplyID) AS TotalItemsSold) to the Supplies table and creates an alternate column name (TotalItemsSold).

The subquery uses the COUNT (*) function to count each record in the Sales table since each record in the Sales table represents a sale.

*Creating Subqueries*

The subquery is executed once for every supply retrieved from the Supplies table.

The outer query retrieves the SupplyID, SupplyName, Price, and SalePrice columns from the Supplies table and sorts the SupplyID column.

Figure 8.13 displays ten records and five columns (SupplyID, SupplyName, Price, SalePrice, TotalItemsSold).

## Results (Output)

| SupplyID | SupplyName | Price | SalePrice | TotalItemsSold |
|---|---|---|---|---|
| AR100 | Animated Rainbow | $20.00 | $18.00 | 2 |
| CC100 | Crystal Cat | $75.00 | $67.50 | 1 |
| CD100 | China Doll | $20.00 | $18.00 | 1 |
| CP100 | China Puppy | $15.00 | $13.50 | 4 |
| DB100 | Dancing Bird | $10.00 | $9.00 | 0 |
| FL100 | Friendly Lion | $14.00 | $12.60 | 1 |
| GR100 | Glass Rabbit | $50.00 | $45.00 | 2 |
| MT100 | Miniature Train Set | $60.00 | $54.00 | 2 |
| PS100 | Praying Statue | $25.00 | $22.50 | 1 |
| WC100 | Wooden Clock | $11.00 | $9.90 | 3 |

**Figure 8.13**

When creating subqueries there is usually more than one way to achieve the same result.

## Conclusion

In this chapter, you learned how to create correlated and non-correlated subqueries. You also learned how to link queries to

Chapter 8

subqueries using the IN, EXISTS, ANY and ALL keywords and how to nest subqueries in queries.

## Test Your Knowledge of the Chapter

## Quiz 8

1. True or False: A subquery is a combination of two or more queries used to produce one result.
2. True or False: Nested subqueries are processed beginning with the innermost SELECT statement.
3. True or False: Subqueries should be surrounded in brackets.
4. True or False: The IN keyword can be used to specify a subquery.
5. True or False: Aggregate functions cannot be used in subqueries.

## Assignment 8

Use the Customers and Sales table in figure 8.8 and 8.9 to retrieve the customer ID, first name and last name of every customer that purchased the China Doll (CD100).

# Chapter 9

# Creating Views

## Introduction

In this chapter, you will learn how to create a view. You will learn how to create a view on one or more tables and how to filter, update and insert data into tables using views. You will also learn how to delete a view. Read the important terms for this chapter.

## Important Terms:

**CREATE VIEW**: Used to tell the DBMS to create a new view.

**DROP VIEW**: Used to delete a view.

**SET**: Used to assign a new value to a column.

**UPDATE Keyword**: Used to modify values already stored in the database.

**UPDATE Statement**: Used to update a single row or every row in a table.

**VIEW**: A query stored in memory that queries one or more tables.

The examples in this chapter were created using Microsoft SQL Server.

The syntax for views may differ from one DBMS to another. Check your DBMS documentation for changes in syntax.

Chapter 9

# Views

A *view* is a query stored in memory that queries one or more tables. Views are commonly used to protect data, shorten complex queries, and combine data from multiple tables. They are sometimes referred to as virtual tables because they look like tables and can be referred to in much the same way as tables. Views are not tables but they are used to view and alter data that is stored in tables. Look at the syntax to create a view:

### CREATE VIEW Syntax

CREATE VIEW ViewName AS
SELECT ColumnOne, ColumnTwo, ColumnThree
FROM TableName

The *CREATE VIEW* keywords are used to tell the DBMS to create a new view. After the CREATE VIEW keywords you must specify the name of the view. The AS keyword is used to specify a query on one or more tables.

After you create a view you can use it to retrieve data from one or more tables, filter data, insert data and update data stored in tables. Look at example one which contains a view on one table.

 Microsoft Access does not support views.

## Creating a View on One Table

# Example One

## Supplies Table

| SupplyID | SupplyName | Price | SalePrice | InStock | OnOrder |
|---|---|---|---|---|---|
| AR100 | Animated Rainbow | 20 | 18 | 10 | 20 |
| CC100 | Crystall Cat | 75 | 67.5 | 60 | 20 |
| CD100 | China Doll | 20 | 18 | 200 | 0 |
| CP100 | China Puppy | 15 | 13.5 | 20 | 40 |
| DB100 | Dancing Bird | 10 | 9 | 10 | 20 |
| FL100 | Friendly Lion | 14 | 12.6 | 0 | 30 |
| GR100 | Glass Rabbit | 50 | 45 | 50 | 20 |
| MT100 | Miniature Train Set | 60 | 54 | 1 | 30 |
| PS100 | Praying Statue | 25 | 22.5 | 3 | 40 |
| WC100 | Wooden Clock | 11 | 9.9 | 100 | 0 |

## Figure 9.1

Say you commonly run the following query on the Supplies table in figure 9.1.

SELECT SUM (InStock) AS TotalStock, MAX (Price) AS HighestPrice, MIN (Price) AS LowestPrice, AVG (InStock) AS AverageStockOnItems, COUNT (SupplyName) AS TotalItemsForSale
FROM Supplies;

Figure 9.2 shows the results from the preceding query.

## Results (Output)

| TotalStock | HighestPrice | LowestPrice | AverageStockOnItems | TotalItemsForSale |
|---|---|---|---|---|
| 454 | 75.00 | 10.00 | 45 | 10 |

## Figure 9.2

Instead of continually retyping this query you could create a view to store the query for future use. Look at the following script that creates a view named PriceStatistics to store the preceding query.

```
CREATE VIEW PriceStatistics AS
SELECT SUM (InStock) AS TotalStock, MAX (Price) AS
HighestPrice, MIN (Price) AS LowestPrice, AVG (InStock) AS
AverageStockOnItems, COUNT (SupplyName) AS
TotalItemsForSale
FROM Supplies
```

In the preceding script, the CREATE VIEW keywords tell the DBMS to create a new view named PriceStatistics. The AS keyword specifies the query on the Supplies table.

Now the query is stored in the computer as a view named PriceStatistics. The view enables you to query it like a table. For example, to view all the records stored in the PriceStatistics view simply type the following:

```
SELECT *
FROM PriceStatistics;
```

The preceding script retrieves the same result as the original query on the Supplies table. Figure 9.3 shows the results from the query.

## Results (Output)

| TotalStock | HighestPrice | LowestPrice | AverageStockOnItems | TotalItemsForSale |
|---|---|---|---|---|
| 454 | 75.00 | 10.00 | 45 | 10 |

### Figure 9.3

NOTE: When you change data in tables that are contained in a view the output of the view also changes.

>  SQL Server does not require a semi colon after a view.

## Creating a View on Multiple Tables

# Example 2

Views are perfect for storing complex queries. This example creates a view to store a complex join on the Supplies, Customers, and Sales tables in figure 9.4, 9.5 and 9.6.

## Supplies Table

| SupplyID | SupplyName | Price | SalePrice | InStock | OnOrder |
|----------|-------------------|-------|-----------|---------|---------|
| AR100 | Animated Rainbow | 20 | 18 | 10 | 20 |
| CC100 | Crystall Cat | 75 | 67.5 | 60 | 20 |
| CD100 | China Doll | 20 | 18 | 200 | 0 |
| CP100 | China Puppy | 15 | 13.5 | 20 | 40 |
| DB100 | Dancing Bird | 10 | 9 | 10 | 20 |
| FL100 | Friendly Lion | 14 | 12.6 | 0 | 30 |
| GR100 | Glass Rabbit | 50 | 45 | 50 | 20 |
| MT100 | Miniature Train Set | 60 | 54 | 1 | 30 |
| PS100 | Praying Statue | 25 | 22.5 | 3 | 40 |
| WC100 | Wooden Clock | 11 | 9.9 | 100 | 0 |

**Figure 9.4**

Chapter 9

## Customers Table

| CustomerID | Firstname | Lastname | Address | City | State | Zipcode | Areacode | PhoneNumber |
|---|---|---|---|---|---|---|---|---|
| 1 | Tom | Evans | 3000 2nd Ave S | Atlanta | GA | 98718 | 301 | 232-9000 |
| 2 | Larry | Genes | 1100 23rd Ave S | Tampa | FL | 33618 | 813 | 982-3455 |
| 3 | Sherry | Jones | 100 Free St S | Tampa | FL | 33618 | 813 | 890-4231 |
| 4 | April | Jones | 2110 10th St S | Santa Fe | NM | 88330 | 505 | 434-1111 |
| 5 | Jerry | Jones | 798 22nd Ave S | St. Pete | FL | 33711 | 727 | 327-3323 |
| 6 | John | Little | 1500 Upside Loop N | St. Pete | FL | 33711 | 727 | 346-1234 |
| 7 | Gerry | Lexingtion | 5642 5th Ave S | Atlanta | GA | 98718 | 301 | 832-8912 |
| 8 | Henry | Denver | 8790 8th St N | Holloman | NM | 88330 | 505 | 423-8900 |
| 9 | Nancy | Kinn | 4000 22nd St S | Atlanta | GA | 98718 | 301 | 879-2345 |
| 10 | Derick | Penns | 2609 15th Ave N | Tampa | FL | 33611 | 813 | 346-1232 |

**Figure 9.5**

## Sales Table

| SalesID | SupplyID | CustomerID | DateSold |
|---|---|---|---|
| 1 | AR100 | 2 | 2/3/2003 |
| 2 | WC100 | 8 | 2/5/2003 |
| 3 | AR100 | 7 | 2/6/2003 |
| 4 | FL100 | 1 | 2/8/2003 |
| 5 | MT100 | 3 | 2/8/2003 |
| 6 | GR100 | 4 | 2/10/2003 |
| 7 | WC100 | 5 | 2/22/2003 |
| 8 | PS100 | 9 | 2/20/2003 |
| 9 | CD100 | 6 | 2/18/2003 |
| 10 | CP100 | 10 | 2/17/2003 |
| 11 | CP100 | 10 | 2/17/2003 |
| 12 | CP100 | 5 | 2/17/2003 |
| 13 | CC100 | 4 | 2/17/2003 |
| 14 | GR100 | 3 | 2/8/2003 |
| 15 | MT100 | 2 | 2/17/2003 |
| 16 | WC100 | 1 | 2/8/2003 |
| 17 | CP100 | 3 | 2/8/2003 |

**Figure 9.6**

*Creating Views*

The following script creates a natural join on three tables (Sales, Customers and Supplies). It displays information on customer purchases. It retrieves every column from the Sales tables, the Firstname and Lastname columns from the Customers table, and the SupplyName column from the Supplies table.

SELECT S.*, C.Firstname, C.Lastname, SP.SupplyName
FROM Sales AS S, Customers AS C, Supplies AS SP
WHERE S.CustomerID = C.CustomerID
AND S.SupplyID = SP.SupplyID;

To create a view to store the preceding natural join on the Sales, Customers and Supplies tables type the following script:

CREATE VIEW CustomerSalesStats AS
SELECT S.*, C.Firstname, C.Lastname, SP.SupplyName
FROM Sales AS S, Customers AS C, Supplies AS SP
WHERE S.CustomerID = C.CustomerID
AND S.SupplyID = SP.SupplyID

The preceding script uses the CREATE VIEW keywords tell to the DBMS to create a new view named CustomerSalesStats. The AS keyword specifies the query on the Sales, Customers and Supplies table.

To view all the records stored in CustomerSalesStats view simply type the following:

SELECT *
FROM CustomerSalesStats;

The preceding script retrieves the same result as the original query on the Sales, Customers and Supplies table. Figure 9.7 shows the results from the query.

Chapter 9

## Results (Output)

| SalesID | SupplyID | CustomerID | DateSold | Firstname | Lastname | SupplyName |
|---|---|---|---|---|---|---|
| 1 | AR100 | 2 | 2/3/2003 | Larry | Genes | Animated Rainbow |
| 2 | WC100 | 8 | 2/5/2003 | Henry | Denver | Wooden Clock |
| 3 | AR100 | 7 | 2/6/2003 | Gerry | Lexington | Animated Rainbow |
| 4 | FL100 | 1 | 2/8/2003 | Tom | Evans | Friendly Lion |
| 5 | MT100 | 3 | 2/8/2003 | Sherry | Lee | Miniature Train Set |
| 6 | GR100 | 4 | 2/10/2003 | April | Lee | Glass Rabbit |
| 7 | WC100 | 5 | 2/22/2003 | Jerry | Lee | Wooden Clock |
| 8 | PS100 | 9 | 2/20/2003 | Nancy | Kinn | Praying Statue |
| 9 | CD100 | 6 | 2/18/2003 | John | Little | China Doll |
| 10 | CP100 | 10 | 2/17/2003 | Derick | Penns | China Puppy |
| 11 | CP100 | 10 | 2/17/2003 | Derick | Penns | China Puppy |
| 12 | CP100 | 5 | 2/17/2003 | Jerry | Lee | China Puppy |
| 13 | CC100 | 4 | 2/17/2003 | April | Lee | Crystal Cat |
| 14 | GR100 | 3 | 2/8/2003 | Sherry | Lee | Glass Rabbit |
| 15 | MT100 | 2 | 2/17/2003 | Larry | Genes | Miniature Train Set |
| 16 | WC100 | 1 | 2/8/2003 | Tom | Evans | Wooden Clock |
| 17 | CP100 | 3 | 2/8/2003 | Sherry | Lee | China Puppy |

**Figure 9.7**

NOTE: Example one and two retrieves every column from the view, but you can retrieve any column you want from a view just like you would a table.

## Filtering Table Data Using a View

# Example 3

Say you want to use the view (CustomerSalesStats) in figure 9.7 to retrieve the first and last names of every customer that purchased supply ID AR100. Look at the following script.

*Creating Views*

```
SELECT *
FROM CustomerSalesStats
WHERE SupplyID = 'AR100';
```

The preceding script queries the CustomerSalesStats view just like you would a table. The query retrieves every column from the CustomerSalesStats view. The WHERE clause sets a condition to only retrieve records that contain a Supply ID equal to AR100. Figure 9.8 shows the results. As you can see only two records are retrieved.

### Results (Output)

| SalesID | SupplyID | CustomerID | DateSold | Firstname | Lastname | SupplyName |
|---|---|---|---|---|---|---|
| 1 | AR100 | 2 | 2003-02-03... | Larry | Genes | Animated Rainbow |
| 3 | AR100 | 7 | 2003-02-06... | Gerry | Lexingtion | Animated Rainbow |

**Figure 9.8**

In most DBMSs, you can perform almost every operation you perform on a table on a view.

# Updating and Inserting Data into Tables Using Views

## Update a Table Using a View

Views can also be used to update and insert data into tables stored in the database.

To update a table using a view you must use an UPDATE statement. The *UPDATE statement* is used to update a single row or every row in

*Chapter 9*

a table. To create an UPDATE statement you must use the *UPDATE* keyword to modify values already stored in the database and the *SET* keyword to assign a new value to a column. A WHERE clause is also included to set conditions on the data you want to update.

When you update a view every column must belong to the same table and you cannot update a column that is created using an expression or function. Look at example 4.

# Example 4

Say you want to update a name stored in the Customers table. You want to change Nancy Kinn's last name to Freely. You could use the CustomerSalesStats view in figure 9.7 to update the Customers table since the CustomerSalesStats view contains the Lastname column and the CustomerID column from the Customers table. Look at the following script.

UPDATE CustomerSalesStats
SET Lastname = 'Freely'
WHERE Lastname = 'Kinn'
AND CustomerID = 9

In the preceding script, the UPDATE keyword is used to tell the DBMS to modify values stored in the CustomerSalesStats view. The SET keyword is used to tell the DBMS to update the Lastname value to Freely. The WHERE clause sets a condition to update only the last names equal to Kinn with a customer ID equal to nine.

To view the updated Lastname column in the Customers table type the following script:

SELECT *
FROM Customers

Figure 9.9 shows the updated Customers table.

*Creating Views*

## Results (Output)

| CustomerID | Firstname | Lastname | Address | City | State | Zipcode | Areacode | PhoneNumber |
|---|---|---|---|---|---|---|---|---|
| 1 | Tom | Evans | 3000 2nd Ave S | Atlanta | GA | 98718 | 301 | 232-9000 |
| 2 | Larry | Genes | 1100 23rd Ave S | Tampa | FL | 33618 | 813 | 982-3455 |
| 3 | Sherry | Jones | 100 Free St S | Tampa | FL | 33618 | 813 | 890-4231 |
| 4 | April | Jones | 2110 10th St S | Santa Fe | NM | 88330 | 505 | 434-1111 |
| 5 | Jerry | Jones | 798 22nd Ave S | St. Pete | FL | 33711 | 727 | 327-3323 |
| 6 | John | Little | 1500 Upside Loop N | St. Pete | FL | 33711 | 727 | 346-1234 |
| 7 | Gerry | Lexington | 5642 5th Ave S | Atlanta | GA | 98718 | 301 | 832-8912 |
| 8 | Henry | Denver | 8790 8th St N | Holloman | NM | 88330 | 505 | 423-8900 |
| 9 | Nancy | Freely | 4000 22nd St S | Atlanta | GA | 98718 | 301 | 879-2345 |
| 10 | Derick | Penns | 2609 15th Ave N | Tampa | FL | 33611 | 813 | 346-1232 |

**Figure 9.9**

**NOTE** Some DBMSs have restrictions on what data can be updated. Check your DBMS documentation on restrictions.

## Insert Data Into a Table Using a View

## Example 5

Say you want to insert a new customer purchase into the Sales table. You could use the CustomerSalesStats view in figure 9.7 to insert the new purchase since it contains every column from the Sales table. Look at the following script.

INSERT INTO CustomerSalesStats (SalesID, SupplyID, CustomerID, DateSold)
VALUES (18, 'PS100', 4, '3/10/03');

The INSERT INTO keywords are used to specify the name of the view and the columns to insert data into. The VALUES keyword is used to specify the values to insert. Figure 9.10 shows the updated

127

## Chapter 9

Sales table with the new purchase (18, PS100, 4, 3/10/03).

> You must specify the column names when you insert data into a table using a view.

## Results (Output)

| SalesID | SupplyID | CustomerID | DateSold |
|---|---|---|---|
| 1 | AR100 | 2 | 2/3/2003 |
| 2 | WC100 | 8 | 2/5/2003 |
| 3 | AR100 | 7 | 2/6/2003 |
| 4 | FL100 | 1 | 2/8/2003 |
| 5 | MT100 | 3 | 2/8/2003 |
| 6 | GR100 | 4 | 2/10/2003 |
| 7 | WC100 | 5 | 2/22/2003 |
| 8 | PS100 | 9 | 2/20/2003 |
| 9 | CD100 | 6 | 2/18/2003 |
| 10 | CP100 | 10 | 2/17/2003 |
| 11 | CP100 | 10 | 2/17/2003 |
| 12 | CP100 | 5 | 2/17/2003 |
| 13 | CC100 | 4 | 2/17/2003 |
| 14 | GR100 | 3 | 2/8/2003 |
| 15 | MT100 | 2 | 2/17/2003 |
| 16 | WC100 | 1 | 2/8/2003 |
| 17 | CP100 | 3 | 2/8/2003 |
| 18 | PS100 | 4 | 3/10/2003 |

**Figure 9.10**

> When you insert records through a view, every column must belong to the same table.

## Deleting a View

# Example 6

Say you want to delete the CustomerSalesStats view in figure 9.7. Look at the following script.

DROP VIEW CustomerSalesStats

The preceding script uses the DROP VIEW keywords to delete the view named CustomerSalesStats.

# Conclusion

In this chapter, you learned how to create a view. You learned how to create a view on one or multiple tables and how to filter, update and insert data into tables using a view. You also learned how to delete a view.

# Test Your Knowledge of the Chapter

# Quiz 9

1. True or False: A view is a query on one or more tables stored in memory.
2. True or False: To delete a view use the DELETE VIEW keywords.
3. True or False: When you change data in tables contained in a view the output of the view also changes.
4. True or False: The AS keyword is used to assign a new value to a column in an UPDATE statement.
5. True or False: Views are often referred to as virtual tables.

*Chapter 9*

# Assignment 9

Use the Customers table in Figure 9.5 to create a view that retrieves every record from the Customers table. Additionally, write a query to retrieve every record from the view you created.

# Chapter 10

# Advanced Table Creation and Table Management

## Introduction

In this chapter, you will learn how to create tables that contain constraints and indexes. You will also learn how to manage your tables using the ALTER TABLE, DROP TABLE, DROP COLUMN and ADD keywords.

Read the important terms for this chapter.

## Important Terms:

**ADD**: Used to add a column to a table.

**ALTER TABLE**: Used to modify a table.

**Clustered Index**: Displays values physically sorted in ascending order in a table.

**Composite Index**: An index on two or more columns.

**Constraint**: Enables you to control how data is inserted into a table column.

**CREATE INDEX**: Used to instruct the DBMS to create a new index.

**Direct Access Method**: Uses indexes to sort and save values of a column in a different location on the computer.

**DROP COLUMN**: Used to delete a column from a table.

*Chapter 10*

**DROP INDEX**: Used to delete an index.

**DROP TABLE**: Used to delete a table.

**Index**: Sorts and saves the values of a column in a different location on the computer with a pointer pointing to the presorted records.

**Referential Integrity**: Ensures that every record in the database is accurate and reliable. It ensures that a relationship between two tables is valid by ensuring that every foreign key within the database corresponds to a primary key.

**Sequential Access Method**: Searches every record in the database until a match is found.

The examples in this chapter were created using Microsoft SQL Server.

Some DBMSs use slightly different syntax; check your DBMS documentation for changes.

## Table Creation

In chapter two, you learned how to create and populate the Employees table. Now you will learn more on how to control how data is entered into a table and how to improve the performance of a table by implementing constraints and indexes. Look at the following script from chapter two that creates the Employees table.

CREATE TABLE Employees
(

```
SocialSecNum CHAR (11) NOT NULL PRIMARY KEY,
Firstname CHAR (50) NOT NULL,
Lastname CHAR (50) NOT NULL,
Address CHAR (50) NOT NULL,
Zipcode CHAR (10) NOT NULL,
Areacode CHAR (3) NULL,
PhoneNumber CHAR (8) NULL
);
```

The preceding script uses the *CREATE TABLE* keywords to instruct the database management system to create a new table named Employees.

Each column (SocialSecNum, Firstname, Lastname, Address, Zipcode, Areacode and PhoneNumber) contains a datatype and a field size. Remember, the *datatype* specifies the type of data a column can store and the *field size* specifies the maximum number of characters a cell within a column can hold.

The NULL or NOT NULL keyword is specified for each column. The *NULL* keyword indicates that a column can be left blank when entering data in the table and the *NOT NULL* keywords indicate that a column cannot be left blank.

Additionally, the SocialSecNum column is the primary key column.

Figure 10.1 shows the Employees table.

## Employees Table

| SocialSecNum | Firstname | Lastname | Address | Zipcode | Areacode | PhoneNumber |
|---|---|---|---|---|---|---|
| | | | | | | |

## Figure 10.1

>  Some DBMSs may use slightly different syntax; check your DBMS documentation for changes.

# Constraints

*Constraints* enable you to control how data is inserted into a table column. They are important because they help maintain referential integrity. *Referential integrity* ensures that every record in the database is accurate and reliable. It ensures that a relationship between two tables is valid by ensuring that every foreign key within the database corresponds to a primary key. Look at some of the most commonly used types of constraints.

## Commonly Used Constraints

**CHECK**: Used to verify that data meets the criterion set for a column.

**FOREIGN KEY**: Used to enforce referential integrity by linking records in one table to records in another table.

**NOT NULL**: Used to specify that a column cannot be left blank.

**PRIMARY KEY**: Used to uniquely identify every record in a table.

**UNIQUE**: Used to ensure that every value in a column is different.

# Example 1

Say you want to recreate the Employees table in Figure 10.1. This time you want to ensure that each value entered into the zip code column is a number between zero and nine and that each zip code entered is no more than five digits long. Additionally, you want to link the Employees table to another table named Departments. Look

## Advanced Table Creation and Table Management

at the following script, which creates a new Employees and Departments table:

```
CREATE TABLE Employees
(
SocialSecNum CHAR (11) NOT NULL PRIMARY KEY,
Firstname CHAR (50) NOT NULL,
Lastname CHAR (50) NOT NULL,
Address CHAR (50) NOT NULL,
Zipcode NUMERIC (5) NOT NULL CHECK (Zipcode LIKE
'[0-9][0-9][0-9][0-9][0-9]'),
Areacode CHAR (3) NULL,
PhoneNumber CHAR (8) NOT NULL
);
```

The preceding script creates the new Employees table. The new Employees table contains a NUMERIC datatype, a five-digit field size and a CHECK constraint. The NUMERIC datatype enables only numbers to be entered into the Zipcode column. The five-digit field size enables only five characters to be typed in the Zipcode column. The CHECK constraint uses the LIKE operator to set a criterion to accept five numbers that fall between zero and nine. Look at the following script, which creates a Departments table that is linked to the new Employees table.

```
CREATE TABLE Departments
(
DepartmentID CHAR (10) NOT NULL PRIMARY KEY,
SocialSecNum CHAR (11) NOT NULL,
DepartmentName CHAR (30) NOT NULL,
DepartmentPhone CHAR (10) NOT NULL UNIQUE,
FOREIGN KEY (SocialSecNum) REFERENCES Employees
(SocialSecNum)
);
```

The preceding script creates a table named Departments with four columns (DepartmentID, SocialSecNum, DepartmentName, and

Chapter 10

DepartmentPhone). Each column contains a NOT NULL constraint and the DepartmentID column contains a PRIMARY KEY constraint. The DepartmentPhone column uses a UNIQUE constraint to ensure that each phone number entered into the DepartmentPhone column is different. The SocialSecNum column is a foreign key because it links the Departments table to the Employees table. The FOREIGN KEY and REFERENCES keywords are used to link the SocialSecNum column to the SocialSecNum column in the Employees table. Figure 10.2 shows the Departments table.

**Departments Table**

| DepartmentID | SocialSecNum | DepartmentName | DepartmentPhone |
|---|---|---|---|
| | | | |

**Figure 10.2**

 Some DBMSs use slightly different syntax; check your DBMS documentation for changes.

# Indexes

Data is normally retrieved using an access method called Sequential Access Method. The *Sequential Access Method* searches through each and every record in a database until a match is found. This method serves its purpose, but if you have a database that contains many records your access time would be quite slow.

To increase the access time, you can use a method called Direct Access Method. The *Direct Access Method* uses indexes to sort and save values of a column in a different location on the computer. Three commonly used types of indexes are the standard index, composite index, and clustered index.

A standard *index* sorts and saves the values of a column in a different location on the computer with a pointer pointing to the presorted records. The standard index enables you to retrieve records much faster because it searches through records that are already sorted rather than searching through every record in a table. A *composite index* is the same as a standard index except it sorts two or more columns at one time. They are normally used to make an index more selective.

The *clustered index* displays values physically sorted in ascending order in a table. For example, the primary key column of a table is a clustered index because the primary key column of a table is always physically sorted in ascending order. Look at the syntax to create an index.

### CREATE INDEX Syntax

CREATE INDEX IndexName
ON Tablename (ColumnName, ColumnName)

Look at example two, which creates an index on the Customers table.

# Create an Index

# Example 2

Chapter 10

## Customers Table

| CustomerID | Firstname | Lastname | Address | City | State | Zipcode | Areacode | PhoneNumber |
|---|---|---|---|---|---|---|---|---|
| 1 | Tom | Evans | 3000 2nd Ave S | Atlanta | GA | 98718 | 301 | 232-9000 |
| 2 | Larry | Genes | 1100 23rd Ave S | Tampa | FL | 33618 | 813 | 982-3455 |
| 3 | Sherry | Jones | 100 Free St S | Tampa | FL | 33618 | 813 | 890-4231 |
| 4 | April | Jones | 2110 10th St S | Santa Fe | NM | 88330 | 505 | 434-1111 |
| 5 | Jerry | Jones | 798 22nd Ave S | St. Pete | FL | 33711 | 727 | 327-3323 |
| 6 | John | Little | 1500 Upside Loop N | St. Pete | FL | 33711 | 727 | 346-1234 |
| 7 | Gerry | Lexingtion | 5642 5th Ave S | Atlanta | GA | 98718 | 301 | 832-8912 |
| 8 | Henry | Denver | 8790 8th St N | Holloman | NM | 88330 | 505 | 423-8900 |
| 9 | Nancy | Kinn | 4000 22nd St S | Atlanta | GA | 98718 | 301 | 879-2345 |
| 10 | Derick | Penns | 2609 15th Ave N | Tampa | FL | 33611 | 813 | 346-1232 |

**Figure 10.3**

Say you frequently create queries that retrieve values from the Lastname column in the Customers table. You could create an index for the Lastname column to enable the DBMS to retrieve last names faster. Look at the following script.

CREATE INDEX LastnameIndex
ON Customers (Lastname)

The *CREATE INDEX* keywords are used to instruct the DBMS to create a new index named LastnameIndex. The *ON* keyword is used to instruct the DBMS to create the index on the Lastname column from the Customers table.

 Indexes use a lot of memory.

# Delete an Index

## Example 3

To delete the index named LastnameIndex use the *DROP INDEX* keywords. After the DROP INDEX keywords, specify the name of the table and the name of the index, separated by a period. Look at the following script.

DROP INDEX Customers.LastNameIndex

# Table Management

After you create a table you can use a combination of keywords to alter the table. You can add a column using the *ADD* keyword and delete a column using the *DROP COLUMN* keywords. The *ALTER TABLE* keywords must be used with one of the preceding keywords to alter a table. Look at example four, five, and six that show how to alter a table.

# Adding a Column

## Example 4

### Departments Table

| DepartmentID | SocialSecNum | DepartmentName | DepartmentPhone |
|---|---|---|---|
|  |  |  |  |

**Figure 10.4**

Say you want to alter the Departments table in figure 10.4. You want to add a new column named DepartmentHead to the table.

Chapter 10

ALTER TABLE Departments
ADD DepartmentHead CHAR (30);

The preceding script uses the ALTER TABLE keywords to instruct the DBMS to make changes to the Departments table. The ADD keyword is used to add a new column named DepartmentHead to the Departments table. The DepartmentHead column contains a character datatype (CHAR) with a field size of 30. Figure 10.5 shows the Departments table with the new DepartmentHead column.

## New Departments Table

| DepartmentID | SocialSecNum | DepartmentName | DepartmentPhone | DepartmentHead |
|---|---|---|---|---|
| | | | | |

**Figure 10.5**

Microsoft Access uses the ADD COLUMN keywords to add a new column to a table.

Columns are always added to the right of existing columns.

Do not alter columns that contain data.

# Deleting a Column

# Example 5

Say you want to delete the DepartmentHead column from the Departments table. Type the following script:

ALTER TABLE Departments
DROP COLUMN DepartmentHead;

The preceding script uses the ALTER TABLE keywords to instruct the DBMS to make changes to the Departments table. The DROP COLUMN keywords are used to delete the DepartmentHead column from the Departments table. Figure 10.6 shows the Departments table without the DepartmentHead column.

## Results (Output)

| DepartmentID | SocialSecNum | DepartmentName | DepartmentPhone |
|---|---|---|---|
| | | | |

**Figure 10.6**

# Deleting a Table

# Example 6

To delete the Departments table type the following script:

DROP TABLE Departments;

The preceding script uses the *DROP TABLE* keywords to delete the Departments table.

Chapter 10

## Conclusion

In this chapter, you learned how to create a table containing constraints and indexes. You also learned how to alter tables using the ALTER TABLE, DROP TABLE, DROP COLUMN and ADD keywords.

## Test Your Knowledge of the Chapter

## Quiz 10

1. Which constraint is used to verify that data meets the criterion set for a column?
2. Which search method searches every record in the database until a match is found?
3. Which keywords are used to modify a table?
4. Which keywords are used to delete a table?
5. True or False: When you add a column to an existing table, the column is added to the left of the existing columns.

## Assignment 10

Create a table named Books with five columns: BookID, Title, Author, Publisher, and ISBN. Use the NOT NULL constraint on every column and make the BookID column the primary key and the ISBN column unique.

# Chapter 11

# Advanced Topics: Transaction Processing, Stored Procedures, Triggers and Cursors

## Introduction

In this chapter, you will learn how to implement transaction processing, and how to create stored procedures, triggers and cursors.

Read the important terms for this chapter.

## Important Terms:

**BEGIN TRANSACTION**: Used to indicate the beginning of a transaction.

**CLOSE**: Used to close a cursor.

**COMMIT TRANSACTION**: Saves and ends a transaction.

**CREATE PROCEDURE**: Used to instruct the DBMS to create a new procedure.

**CREATE TRIGGER**: Used to instruct the DBMS to create a new trigger.

**CURSOR**: Used to scroll through large amounts of records on the screen.

**DEALLOCATE**: Used to release all memory associated with a cursor.

Chapter 11

**EXECUTE**: Used to run a stored procedure.

**FETCH**: Used to retrieve a row among specified records.

**FETCH NEXT FROM**: Used to instruct the DBMS to locate the first record among specified records.

**IF**: Used to set conditions on values of data retrieved from the database.

**NEXT**: Used to move the cursor to the next row in a result set.

**OPEN**: Used to open a cursor.

**PRINT**: Used to display data to an output device.

**Result set**: Group of records retrieved from an SQL query.

**ROLLBACK**: Used to undo an SQL statement.

**SAVE TRANSACTION**: Used to create savepoints to rollback to within a transaction.

**Savepoint**: Used to hold a place for the DBMS to roll back to if an error occurs.

**Stored Procedure**: One or more SQL statements stored in the database.

**Transaction processing**: Enables you to ensure that a block of SQL statements execute completely or not at all.

**TRANSACTION**: Two or more SQL statements that must be completed together.

**TRIGGER**: A stored procedure that performs specific actions that are linked to specific operations on a single table in the database.

# Advanced Topics: Transaction Processing, Stored Procedures, Triggers and Cursors

The examples in this chapter were created using Microsoft SQL Server.

Some DBMSs use slightly different syntax; check your DBMS documentation for changes.

## Transaction Processing

A *transaction* is two or more SQL statements that must be completed as a group. When you execute a transaction you need to ensure that each and every statement within the transaction processes completely. If not, you may end up with a partial transaction.

Partial transactions take place when errors occur during the processing of an SQL statement. Any of the following can cause a partial transaction: power outage, low memory, computer crash, or a security restriction.

*Transaction processing* helps support referential integrity; it enables you to ensure that any INSERT UPDATE or DELETE statement executes completely or not at all. When implementing transaction processing you create savepoints within a transaction. If an error occurs during the execution of an SQL statement within a transaction, specified SQL statements within the transaction can be undone (rollback) until it reaches a savepoint within the transaction.

In Microsoft SQL Server, all transactions must begin with the *BEGIN TRANSACTION* keywords. The *BEGIN TRANSACTION* keywords are used to indicate the beginning of a new transaction. The *SAVE TRANSACTION* keywords are used to create a savepoint. *Savepoints* hold a place for the DBMS to roll back to if an error occurs. The ROLLBACK TRANSACTION keywords are used to instruct the DBMS to begin undoing SQL statements until it reaches a savepoint. The COMMIT TRANSACTION keywords instruct the DBMS to

Chapter 11

save and end the transaction. Look at example one which
demonstrates transaction processing.

 Microsoft Access does not support transaction processing.

## Create a Transaction

# Example 1

Say you want to run a transaction that inserts a new customer in the
Customers table and a new purchase for the new customer in the Sales
table. You want to ensure that the new customer is entered into the
database even if an error occurs while entering the new order;
therefore you create a savepoint to rollback to if an error occurs. Look
at the following script:

BEGIN TRANSACTION

INSERT INTO Customers (CustomerID, Firstname, Lastname,
Address, City, State, Zipcode, Areacode, PhoneNumber)
VALUES (11, 'Vivian', 'Raines', '1303 13th Ave N', 'Atlanta',
'GA', 98108, 301, '423-8543')
SAVE TRANSACTION NewCustomer
INSERT INTO Sales (SalesID, SupplyID, CustomerID, DateSold)
VALUES (18, 'AR100', 11, '4/5/03')
IF @@ERROR <> 0
BEGIN
PRINT "An error occurred during execution of one of the INSERT
statements."
ROLLBACK TRANSACTION NewCustomer
END

COMMIT TRANSACTION

*Advanced Topics: Transaction Processing, Stored Procedures, Triggers and Cursors*

In the preceding script, the *BEGIN TRANSACTION* keywords are used to indicate the beginning of a new transaction. The INSERT statement inserts a new customer into the Customers table. The *SAVE TRANSACTION* keywords are used to create a savepoint named NewCustomer. The savepoint holds a place for the DBMS to roll back to if an error occurs after the new customer is entered into the Customers table.

Each savepoint must contain a unique name.

The next INSERT statement inserts a new purchase by the new customer into the Sales table.

Variables are used in SQL to store data. The *@@ERROR* variable is a global variable used in Microsoft SQL Server to show that an error occurred. When the @@ERROR variable returns a value other than 0, it means an error occurred. When an error occurs, the transaction rolls back to the first savepoint it comes to.

The IF statement creates a conditional statement. If the @@ERROR variable is not equal to zero, the IF statement instructs the DBMS to print the statement after the PRINT keyword and to use the ROLLBACK TRANSACTION keywords to begin undoing SQL statements until it reaches a savepoint named NewCustomer.

The *BEGIN* and *END* keywords are used to define a series of SQL statements that must be executed together. The BEGIN and END keywords instruct the DBMS to execute the PRINT and ROLLBACK statement together.

The COMMIT TRANSACTION keywords instruct the DBMS to save and end the transaction.

Chapter 11

 **NOTE** Variables differ from one DBMS to another and not all DBMSs require the use of the BEGIN TRANSACTION and COMMIT TRANSACTION keywords. Check your DBMS documentation for changes.

## Stored Procedures

A *stored procedure* is one or more SQL statements stored in the database. Stored procedures can be used to store simple SQL statements or complex compiled code like many high-level object-oriented programming languages such as C++ and JAVA.

Every stored procedure must contain its own name. They are executed by typing a simple execute statement containing the EXECUTE keyword and the name of the stored procedure.

Some stored procedures accept input parameters from the user and some execute without input parameters. Input parameters are variables supplied by the user.

When a stored procedure accepts input parameters they are specified after the name of the stored procedure. All parameters must contain a unique name, a datatype, a field size and must start with the @ symbol. Look at the syntax to create a stored procedure.

### CREATE PROCEDURE Syntax

CREATE PROCEDURE ProcedureName | Parameters | Optional
AS
SQL QUERY

The *CREATE PROCEDURE* keywords instruct the DBMS to create a new procedure. The name of the procedure follows the CREATE PROCEDURE keywords. Parameters are specified after the procedure name. After the parameters, you can additionally use control

*Advanced Topics: Transaction Processing, Stored Procedures, Triggers and Cursors*

statements often used in high-level object oriented programming languages. The AS keyword is used to specify SQL statements.

Look at example two, which shows a simple stored procedure that does not accept input parameters.

 Microsoft Access does not support transaction processing.

Chapter 11

# Create a Stored Procedure

## Stored Procedure with no Input Parameters

## Example 2

### Sales Table

| SalesID | SupplyID | CustomerID | DateSold |
|---|---|---|---|
| 1 | AR100 | 2 | 2/3/2003 |
| 2 | WC100 | 8 | 2/5/2003 |
| 3 | AR100 | 7 | 2/6/2003 |
| 4 | FL100 | 1 | 2/8/2003 |
| 5 | MT100 | 3 | 2/8/2003 |
| 6 | GR100 | 4 | 2/10/2003 |
| 7 | WC100 | 5 | 2/22/2003 |
| 8 | PS100 | 9 | 2/20/2003 |
| 9 | CD100 | 6 | 2/18/2003 |
| 10 | CP100 | 10 | 2/17/2003 |
| 11 | CP100 | 10 | 2/17/2003 |
| 12 | CP100 | 5 | 2/17/2003 |
| 13 | CC100 | 4 | 2/17/2003 |
| 14 | GR100 | 3 | 2/8/2003 |
| 15 | MT100 | 2 | 2/17/2003 |
| 16 | WC100 | 1 | 2/8/2003 |
| 17 | CP100 | 3 | 2/8/2003 |

**Figure 11.1**

The following script demonstrates a stored procedure that counts the total amount of purchases for each customer in the Sales table in figure 11.1.

```
CREATE PROCEDURE CountPurchases
AS
SELECT CustomerID, COUNT (*) AS TotalPurchases
FROM Sales
GROUP BY CustomerID
```

*Advanced Topics: Transaction Processing, Stored Procedures, Triggers and Cursors*

The preceding script uses the CREATE PROCEDURE keywords to instruct the DBMS to create a new stored procedure named CountPurchases. The AS keyword is used to specify a query. The query counts each CustomerID in the Sales table and displays the CustomerID column and the total number of purchases per customer ID as TotalPurchases. To execute the stored procedure type the following script:

EXECUTE CountPurchases

The preceding EXECUTE statement is used to run the stored procedure named CountPurchases. Figure 11.2 shows the result from the EXECUTE statement.

## Results (Output)

| CustomerID | TotalPurchases |
|---|---|
| 1 | 2 |
| 2 | 2 |
| 3 | 3 |
| 4 | 2 |
| 5 | 2 |
| 6 | 1 |
| 7 | 1 |
| 8 | 1 |
| 9 | 1 |
| 10 | 2 |

## Figure 11.2

Stored procedures that accept input parameters are more complex. Example three shows a simple stored procedure that accepts an input parameter.

Chapter 11

# Stored Procedure That Accepts an Input Parameter

# Example 3

## Supplies Table

| SupplyID | SupplyName | Price | SalePrice | InStock | OnOrder |
|---|---|---|---|---|---|
| AR100 | Animated Rainbow | 20 | 18 | 10 | 20 |
| CC100 | Crystall Cat | 75 | 67.5 | 60 | 20 |
| CD100 | China Doll | 20 | 18 | 200 | 0 |
| CP100 | China Puppy | 15 | 13.5 | 20 | 40 |
| DB100 | Dancing Bird | 10 | 9 | 10 | 20 |
| FL100 | Friendly Lion | 14 | 12.6 | 0 | 30 |
| GR100 | Glass Rabbit | 50 | 45 | 50 | 20 |
| MT100 | Miniature Train Set | 60 | 54 | 1 | 30 |
| PS100 | Praying Statue | 25 | 22.5 | 3 | 40 |
| WC100 | Wooden Clock | 11 | 9.9 | 100 | 0 |

**Figure 11.3**

## Sales Table

| SalesID | SupplyID | CustomerID | DateSold |
|---|---|---|---|
| 1 | AR100 | 2 | 2/3/2003 |
| 2 | WC100 | 8 | 2/5/2003 |
| 3 | AR100 | 7 | 2/6/2003 |
| 4 | FL100 | 1 | 2/8/2003 |
| 5 | MT100 | 3 | 2/8/2003 |
| 6 | GR100 | 4 | 2/10/2003 |
| 7 | WC100 | 5 | 2/22/2003 |
| 8 | PS100 | 9 | 2/20/2003 |
| 9 | CD100 | 6 | 2/18/2003 |
| 10 | CP100 | 10 | 2/17/2003 |
| 11 | CP100 | 10 | 2/17/2003 |
| 12 | CP100 | 5 | 2/17/2003 |
| 13 | CC100 | 4 | 2/17/2003 |
| 14 | GR100 | 3 | 2/8/2003 |
| 15 | MT100 | 2 | 2/17/2003 |
| 16 | WC100 | 1 | 2/8/2003 |
| 17 | CP100 | 3 | 2/8/2003 |

## Figure 11.4

The following script creates a stored procedure that accepts an input parameter to retrieve sales data from the Supplies and Sales table in figure 11.3 and 11.4.

CREATE PROCEDURE SalesStatistics
@splyid varchar (5)
AS
SELECT Sales.CustomerID, Sales.SupplyID, Supplies.SupplyName, Supplies.Price, Supplies.SalePrice, Sales.DateSold
FROM Supplies, Sales
WHERE Supplies.SupplyID = Sales.SupplyID
AND Sales.SupplyID = @splyid

The preceding script uses the CREATE PROCEDURE keywords to instruct the DBMS to create a new stored procedure named

Chapter 11

SalesStatistics. The @splyid is an input parameter. The input parameter is used to instruct the DBMS to expect input from the user when the stored procedure is executed. The input parameter is defined as a varchar datatype with a field size of five. This means the user must input a varchar value no more than five characters long.

> All parameters must contain a unique name, a datatype, a field size and must begin with the @ symbol.

The AS keyword is used to specify an inner join. The inner join retrieves columns from the Supplies and Sales table. The WHERE clause shows the relationship between the Supplies and Sales table. The WHERE clause also sets the @splyid parameter equal to the SupplyID column in the Sales table. Setting the @splyid parameter equal to the SupplyID column in the Sales table causes the DBMS to expect a value equal to one of the supply ID values stored in the Sales table.

To execute the stored procedure, type the EXECUTE keyword along with the name of the stored procedure and a supply ID. Type the following script to retrieve sales data on supply ID CP100:

Execute SalesStatistics 'CP100'

Figure 11.5 shows the result from the EXECUTE statement.

## Results (Output)

| CustomerID | SupplyID | SupplyName | Price | SalePrice | DateSold |
|---|---|---|---|---|---|
| 10 | CP100 | China Puppy | 15.00 | 13.50 | 2003-02-17 00:00:00.000 |
| 10 | CP100 | China Puppy | 15.00 | 13.50 | 2003-02-17 00:00:00.000 |
| 5 | CP100 | China Puppy | 15.00 | 13.50 | 2003-02-17 00:00:00.000 |
| 3 | CP100 | China Puppy | 15.00 | 13.50 | 2003-02-08 00:00:00.000 |

## Figure 11.5

*Advanced Topics: Transaction Processing, Stored Procedures, Triggers and Cursors*

 Some DBMSs use slightly different syntax; check your DBMS documentation for changes.

## Delete a Stored Procedure

To delete a stored procedure, type the DROP PROCEDURE keywords followed by the name of the stored procedure. To delete the SalesStatistics procedure, type the following script:

DROP PROCEDURE SalesStatistics

## Triggers

Another type of stored procedure is called a trigger. *Triggers* perform specific actions that are linked to specific operations on a single table in the database. They execute automatically either before or after an UPDATE, INSERT or DELETE statement.

 Triggers cannot include SELECT statements.

Triggers are often used to check the integrity of the database, update special fields and enforce security. Look at the syntax to create a trigger.

### CREATE TRIGGER Syntax

CREATE TRIGGER TriggerName
ON TableName
FOR INSERT | UPDATE | DELETE
AS Operation

Chapter 11

The *CREATE TRIGGER* keywords instruct the DBMS to create a new trigger. The name of the trigger follows the CREATE TRIGGER keywords. The ON keyword is used to specify the table the trigger is linked to, and the FOR keyword is used to specify the operation that the trigger is linked to. The AS keyword is used to specify the operation the trigger will carry out.

## Create a Trigger

When you create a trigger, you must specify specific conditions in which the trigger will execute, and the action the trigger must carry out. Look at example four which shows a trigger.

## Example 4

Say you want to display a reminder message whenever a new supply is inserted into the Supplies table or whenever an existing supply is updated in the Supplies table. Look at the following script:

CREATE TRIGGER ReminderForNewSupplies
ON Supplies
FOR INSERT, UPDATE
AS PRINT "All sale prices for new supplies are 10% off unless otherwise noted."

The preceding script uses the CREATE TRIGGER keywords to instruct the DBMS to create a new trigger named ReminderForNewSupplies. The ON keyword is used to link the trigger to the Supplies table. The FOR keyword specifies the operation (INSERT, UPDATE) the trigger is linked to. The AS keyword specifies the operation (print) the trigger will carry out.

Whenever an INSERT or UPDATE statement is executed on the Supplies table the trigger will execute the print statement after the PRINT keyword. Look at the following INSERT statement.

*Advanced Topics: Transaction Processing, Stored Procedures,
Triggers and Cursors*

## INSERT Statement

INSERT INTO Supplies
VALUES ('LD100', 'Leather Deer', 20.00, 18.00, 200, 0);

The preceding INSERT statement inserts a new record into the Supplies table and causes the following message to display on the screen.

## Message

All sale prices for new supplies are 10% off unless otherwise noted.

Figure 11.6 shows the message.

## Results (Output)

```
All sale prices for new supplies are 10% off unless otherwise noted.
(1 row(s) affected)
```

## Figure 11.6

In Microsoft SQL Server, click on the message tab near the bottom of the screen to see the message.

Some DBMSs use slightly different syntax; check your DBMS documentation for changes.

Chapter 11

## Delete a Trigger

To delete a trigger, type the DROP TRIGGER keywords and the name of the trigger. To delete the ReminderForNewSupplies trigger, type the following statement.

DROP TRIGGER ReminderForNewSupplies

## Cursors

*Cursors* are used to scroll through large amounts of records on the screen. When you create queries that retrieve many records you can use a cursor to scroll through records one at a time. Cursors are also used to save retrieved records for future use.

The DECLARE and CURSOR keywords are used to create and name a new cursor. These keywords also instruct the DBMS to allocate space for a new cursor. Cursors have to be open in order to use them. When you finish using a cursor you must close it. When you open a cursor it retrieves the records you specified for scrolling. Use the OPEN CURSOR keywords to open a cursor and the CLOSE keyword to close a cursor.

The *FETCH* keyword is used to retrieve a row among specified records. When you finish using a cursor use the *DEALLOCATE* keyword to release all memory associated with the cursor.

The *NEXT* keyword is used to move the cursor to the next row in a result set. A *result set* is a group of records retrieved from an SQL query. The FROM keyword is used to specify the cursor name that the fetch refers to.

Global variables are often used in cursors. The @@FETCH_STATUS variable is a global variable used in SQL Server to test a cursors fetch status. If the @@FETCH_STATUS variable returns a 0, the FETCH statement was successful. If it returns a -1, the FETCH statement failed, and if it returns a -2, the row that was fetched is missing. All global variables begin with the @@ symbol. Look at example five, which shows a cursor.

 You cannot use a cursor that is closed. You must reopen it.

## Create a Cursor

## Example 5

Say you want to create a cursor to scroll records in the Employees table in figure 11.7. You want to scroll all the names and phone numbers of employees that have a zip code equal to or between 33612 and 98800.

### Employees Table

| SocialSecNum | Firstname | Lastname | Address | Zipcode | Areacode | PhoneNumber |
|---|---|---|---|---|---|---|
| 109-83-4765 | Shaun | Rivers | 1548 6th Ave S Atlanta, GA | 98718 | 301 | 894-1973 |
| 123-88-1982 | Debra | Fields | 1934 16th Ave N Atlanta, GA | 98718 | 301 | 897-3245 |
| 211-73-1112 | Tom | Jetson | 1311 2nd Ave E Atlanta, GA | 98718 | 301 | 897-9877 |
| 226-73-1919 | Jacob | Lincoln | 2609 40th Ave S Honolulu, HI | 96820 | 808 | 423-4111 |
| 249-74-1682 | Jackie | Fields | 2211 Peachtree St N Tampa, FL | 33612 | 813 | 827-2301 |
| 263-73-1442 | Adam | Williams | 1938 32nd Ave S. St. Pete, FL | 33711 | 727 | 321-2234 |
| 266-11-4444 | Sam | Elliot | 1601 Center Loop Tampa, FL | 33612 | 813 | 898-2134 |
| 266-73-1982 | John | Dentins | 2211 22nd Ave N Atlanta, GA | 98718 | 301 | 897-4321 |
| 980-22-1982 | Shawn | Lewis | 1601 4th Ave W Atlanta, GA | 98718 | 301 | 894-0987 |
| 982-24-3490 | Yolanda | Brown | 1544 16th Ave W Atlanta, GA | 98718 | 301 | 892-1234 |

## Figure 11.7

DECLARE CustZip3361298800 CURSOR
FOR
SELECT Firstname, Lastname, PhoneNumber
FROM Employees

```
WHERE Zipcode BETWEEN 33612 AND 98800
OPEN CustZip3361298800
FETCH NEXT FROM CustZip3361298800
WHILE @@FETCH_STATUS = 0
BEGIN
FETCH NEXT FROM CustZip3361298800
END
CLOSE CustZip3361298800
DEALLOCATE CustZip3361298800
```

The preceding script uses the DECLARE CURSOR keywords to instruct the DBMS to create a new cursor. The name (CustZip3361298800) of the cursor is specified between the DECLARE and CURSOR keywords.

The FOR keyword is used to specify a SELECT statement that retrieves the name and phone number of employees that have a zip code equal to or between 33612 and 98800.

The OPEN keyword is used to open the CustZip3361298800 cursor. The *FETCH NEXT FROM* keywords instruct the DBMS to locate the first record from the SELECT statement. The WHILE statement is a conditional statement that uses the BEGIN FETCH NEXT FROM END keywords to instruct the DBMS to continue scrolling all the records as long as the @@FETCH_STATUS global variable is equal to zero.

The CLOSE keyword is used to close the curser and the DEALLOCATE keyword is used to release all memory associated with the cursor.

Some DBMSs use slightly different syntax; check your DBMS documentation for changes.

# Test Your Knowledge of the Chapter

# Quiz 11

1. True or False: The CREATE PROCEDURE keywords are used to create a stored procedure.
2. True or False: Savepoints are used to display data to an output device.
3. True or False: The DEALLOCATE keyword is used to release all memory associated with the cursor.
4. True or False: A result set is a group of records retrieved from an SQL query.
5. What type of procedure is linked to specific operations on a single table in the database?

# Assignment 11

Create a stored procedure and an execute statement to retrieve all the records from the employees table in figure 11.7.

# Appendix A

# Answers to Quizzes and Assignments

This appendix provides answers to the quizzes and assignments throughout the chapters of the book.

# Quiz 1

1. True or False: A column is a record within a table that runs horizontally within a table. FALSE
2. True or False: Microsoft created Structured Query Language. FALSE
3. True or False: A primary key is a field whose value uniquely identifies every row in a table. TRUE
4. True or False: Normalization is a technique used to organize data attributes in a more efficient, reliable, flexible and maintainable structure. TRUE
5. What links records of one type to those of another type? FOREIGN KEY

# Assignment 1

Create columns for two tables on paper. Link the two tables by assigning a primary key to both tables and a foreign key to one table.

Students Table: *StudentID*, Name, Address, Zipcode, PhoneNumber

Courses Table: *CourseID*, *StudentID*, Course, StartTime, EndTime, StartDate, EndDate, Teacher, Credit

*Answers to Quizzes and Assignments*

# Quiz 2

1. What symbol is used to select every column in a table? The asterisk (*)
2. True or False: The delete statement can be used to delete tables. FALSE
3. True or False: A field is equivalent to a column. TRUE
4. True or False: *NOT NULL* indicates that a field can be left blank when entering data into a table. FALSE
5. True or False: The SELECT INTO keywords are used to transfer data to an existing table. FALSE

# Assignment 2

Create and populate a table with three records.

CREATE TABLE Words (
WordID CHAR (50) Primary Key NOT NULL,
WordOne CHAR (50) NOT NULL,
WordTwo CHAR (50) NOT NULL
);

INSERT INTO Words
Values ('See', 'Spot', 'Run');
INSERT INTO Words
Values ('You', 'Are', 'Great');
INSERT INTO Words
Values ('Keep', 'Them', 'Coming');

# Quiz 3

1. What keyword is used to create an alternate name for a column? AS
2. What symbol is used to perform concatenation? (+)
3. True or False: UNION ALL is used to combine two queries to show all duplicates records. TRUE

*Appendix A*

4. True or False: Some DBMSs use the (||) symbol in place of the (+) symbol to perform concatenation. TRUE
5. True or False: The FROM keyword is used to specify specific column(s) in a table. FALSE

## Assignment 3

Use the Employees table from this chapter to create a query that retrieves the SocialSecNum, Firstname, and Lastname columns. Merge the Firstname and Lastname columns and create an alternate column name for the merged columns.

SELECT SocialSecNum, Firstname + Lastname AS FullName FROM Employees;

## Quiz 4

1. Which character operator is used with the percent symbol to match parts of a value? LIKE
2. Which type of operator is used to perform wildcard-character searches? Character
3. True or False: The WHERE clause is used to combine two queries to show all duplicates records. FALSE
4. True or False: Operators are used in the WHERE clause to set conditions on data. TRUE
5. Which type of operator is used to separate two or more conditions in a WHERE clause? Logical

## Assignment 4

Use the Courses table in figure 4.7 to create a query that shows the courses for students with the following student ID's: 1, 2 and 3

```
SELECT *
FROM Courses
WHERE StudentID IN (1, 2, 3);
```

# Quiz 5

1. Which arithmetic operator is used to perform multiplication? (*)
2. What function is used to display the system time and date? GETDATE ()
3. True or False: The SUM () function counts the number of rows in a column. FALSE
4. True or False: The COUNT () function returns the sum of a column. FALSE
5. What function replaces specified data? REPLACE ()

# Assignment 5

Use the Numbers table in figure 5.7 to create a query that sums Column 1, averages Column 2, and counts Column 3. Display the results using alternate column names.

SELECT SUM (Column1) AS SumColumn1, AVG (Column2) AS AVGColumn2, COUNT (Column3) AS COUNTColumn3
FROM Numbers;

# Quiz 6

1. What keyword is used to sort a column in descending order? DESC
2. What clause is used to sort groups of data calculated from aggregate functions? GROUP BY
3. True or False: You must use the ASC keyword to sort columns in ascending order. FALSE
4. True or False: Aggregate functions can be used in the WHERE clause. FALSE

*Appendix A*

5. True or False: The ORDER BY clause is used to sort specified columns in ascending or descending order. TRUE

## Assignment 6

Use the Sales table in figure 6.12 to create a query that locates the first order date for each customer. Group the results by the CustomerID column.

SELECT CustomerID, MIN (DateSold) AS FirstOrderDate
FROM Sales
GROUP BY CustomerID;

## Quiz 7

1. True or False: Joins enable you to use multiple SELECT statements to query two or more tables at the same time. FALSE
2. True or False: An Inner Join matches values of a column in one table to matching values in the same table. FALSE
3. Name the two types of outer joins. LEFT OUTER JOIN and RIGHT OUTER JOIN
4. Which type of join is used to specify only unique columns from multiple tables? Natural join
5. Which keyword works like the WHERE clause and is used with the OUTER JOIN keywords? ON

## Assignment 7

Use the Customers table in figure 7.1 and the Sales table in figure 7.2 to create an inner join displaying the Lastname and SalesID columns.

SELECT Lastname, SalesID
FROM Customers, Sales
WHERE Customers.CustomerID = Sales.CustomerID;

*Answers to Quizzes and Assignments*

# Quiz 8

1. True or False: A subquery is a combination of two or more queries used to produce one result. FALSE
2. True or False: Nested subqueries are processed beginning with the innermost SELECT statement. TRUE
3. True or False: Subqueries should be surrounded in brackets. FALSE
4. True or False: The IN keyword can be used to specify a subquery. TRUE
5. True or False: Aggregate functions cannot be used in subqueries. FALSE

# Assignment 8

Use the Customers and Sales table in figure 8.8 and 8.9 to retrieve the customer ID, first name and last name of every customer that purchased the China Doll (CD100).

SELECT CustomerID, Firstname, Lastname
FROM Customers
WHERE CustomerID

IN

(SELECT CustomerID
FROM Sales
WHERE SupplyID = 'CD100');

# Quiz 9

1. True or False: A view is a query on one or more tables stored in memory. TRUE

*Appendix A*

2. True or False: To delete a view use the DELETE VIEW keywords. FALSE
3. True or False: When you change data in tables contained in a view the output of the view also changes. TRUE
4. True or False: The AS keyword is used to assign a new value to a column in an UPDATE statement. FALSE
5. True or False: Views are often referred to as virtual tables. TRUE

## Assignment 9

Use the Customers table in Figure 9.5 to create a view that retrieves every record from the Customers table. Additionally, write a query to retrieve every record from the view you created.

CREATE VIEW CustomerRecords AS
SELECT *
FROM Customers

SELECT *
FROM CustomerRecords;

## Quiz 10

1. Which constraint is used to verify that data meets the criterion set for a column? CHECK
2. Which search method searches every record in the database until a match is found? Sequential Access Method
3. Which keywords are used to modify a table? ALTER TABLE
4. Which keywords are used to delete a table? DROP TABLE
5. True or False: When you add a column to an existing table, the column is added to the left of the existing columns. FALSE

# Assignment 10

Create a table named Books with five columns: BookID, Title, Author, Publisher, and ISBN. Use the NOT NULL constraint on every column and make the BookID column the primary key and the ISBN column unique.

CREATE TABLE Books
(
BookID CHAR (10) NOT NULL PRIMARY KEY,
Title CHAR (50) NOT NULL,
Author CHAR (30) NOT NULL,
Publisher CHAR (30) NOT NULL,
ISBN CHAR (13) NOT NULL UNIQUE,
);

# Quiz 11

1. True or False: The CREATE PROCEDURE keywords are used to create a stored procedure. TRUE
2. True or False: Savepoints are used to display data to an output device. FALSE
3. True or False: The DEALLOCATE keyword is used to release all memory associated with the cursor. TRUE
4. True or False: A result set is a group of records retrieved from an SQL query. TRUE
5. What type of procedure is linked to specific operations on a single table in the database? Trigger

# Assignment 11

Create a stored procedure and an execute statement to retrieve all the records from the employees table in figure 11.7.

*Appendix A*

CREATE PROCEDURE EmployeeInfo
AS
SELECT *
FROM Employees
ORDER BY Lastname

EXECUTE EmployeeInfo

# Appendix B

# SQL Script for the Tables Used in the Book

This appendix provides the SQL script to create and populate the tables used throughout the book.

Use the CREATE TABLE script to create the tables and the INSERT statements to populate the tables.

This script is also available on-line at:

http://www.jaffainc.com/SQLBook

The script is written in the simplest form to accommodate multiple Database Management Systems (DBMSs). Check your DBMS documentation for changes. Before you begin make sure you read the important notes.

Do not run the INSERT Statements more than once or it may generate errors.

In some DBMSs you may have to set your primary key as follows: CustomerID INTEGER NOT NULL CONSTRAINT PriKey Primary Key,

In Access, when you do not state NOT NULL when creating a column, the column is automatically set to NULL. However, in some DBMSs, if you want to set a column to NULL you must explicitly state NULL.

*Appendix B*

 In some DBMSs you may need to substitute a DECIMAL (8, 2), or CURRENCY datatype for the MONEY datatype in the Create Table script.

 Oracle users must use the DATE datatype in place of the DATETIME datatype in the Create Table script.

## Committee1 Table

```
CREATE TABLE Committee1
(
SocialSecNum CHAR (11) NOT NULL PRIMARY KEY,
Firstname CHAR (50) NOT NULL,
Lastname CHAR (50) NOT NULL,
Address CHAR (50) NOT NULL,
Zipcode CHAR (10) NOT NULL,
Areacode CHAR (3) NULL,
PhoneNumber CHAR (8) NULL
);
```

## Insert Statements to Populate the Committee1 Table

INSERT INTO Committee1
VALUES ('263-73-1442', 'Adam', 'Williams', '1938 32nd Ave S. St. Pete, FL', 33711, 727, '321-2234');

INSERT INTO Committee1
VALUES ('226-73-1919', 'Jacob', 'Lincoln', '2609 40th Ave S Honolulu, HI', 96820, 808, '423-4111');

INSERT INTO Committee1
VALUES ('249-74-1682', 'Jackie', 'Fields', '2211 Peachtree St N Tampa, FL', 33612, 813, '827-2301');

INSERT INTO Committee1
VALUES ('123-88-1982', 'Debra', 'Fields', '1934 16th Ave N Atlanta, GA', 98718, 301, '897-3245');

INSERT INTO Committee1
VALUES ('211-73-1112', 'Tom', 'Jetson', '1311 2nd Ave E Atlanta, GA', 98718, 301, '897-9877');

## Committee2 Table

CREATE TABLE Committee2
(
SocialSecNum CHAR (11) NOT NULL PRIMARY KEY,
Firstname CHAR (50) NOT NULL,
Lastname CHAR (50) NOT NULL,
Address CHAR (50) NOT NULL,
Zipcode CHAR (10) NOT NULL,
Areacode CHAR (3) NULL,
PhoneNumber CHAR (8) NULL
);

## Insert Statements to Populate the Committee2 Table

INSERT INTO Committee2
VALUES ('266-73-1982', 'John', 'Dentins', '2211 22nd Ave N Atlanta, GA', 98718, 301, '897-4321');

INSERT INTO Committee2
VALUES ('266-11-4444', 'Sam', 'Elliot', '1601 Center Loop Tampa, FL', 33612, 813, '898-2134');

INSERT INTO Committee2
VALUES ('263-73-1442', 'Adam', 'Williams', '1938 32nd Ave S. St. Pete, FL', 33711, 727, '321-2234');

*Appendix B*

INSERT INTO Committee2
VALUES ('226-73-1919', 'Jacob', 'Lincoln', '2609 40th Ave S Honolulu, HI', 96820, 808, '423-4111');

INSERT INTO Committee2
VALUES ('249-74-1682', 'Jackie', 'Fields', '2211 Peachtree St N Tampa, FL', 33612, 813, '827-2301');

## Courses Table

CREATE TABLE Courses
(
CourseID CHAR (20) NOT NULL Primary Key,
StudentID CHAR (4) NOT NULL,
Course CHAR (50) NOT NULL,
StartTime CHAR (50) NOT NULL,
EndTime CHAR (50) NOT NULL,
StartDate DATETIME NOT NULL,
EndDate DATETIME NOT NULL,
Teacher CHAR (30) NOT NULL,
Credit CHAR (2) NOT NULL
);

## Insert Statements to Populate the Courses Table

INSERT INTO Courses
Values ('M1101', 1, 'Pre Algebra', '3:00pm', '5:00pm', '2/3/03', '5/3/03', 'Mr. Stevens', 3);

INSERT INTO Courses
Values ('M1102', 5, 'Pre Calculus', '3:00pm', '5:00pm', '2/3/03', '5/3/03', 'Mr. Dixon', 3);

INSERT INTO Courses
Values ('L1001', 3, 'Literature', '2:00pm', '4:00pm', '2/3/03', '5/3/03', 'Mrs. Donaldson', 3);

## SQL Script for the Tables Used in the Book

INSERT INTO Courses
Values ('R1001', 2, 'Reading', '1:00pm', '3:00pm', '2/3/03',
'5/3/03', 'Ms Jackson', 3);

INSERT INTO Courses
Values ('M1103', 4, 'Statistics', '3:00pm', '5:00pm', '2/3/03',
'5/3/03', 'Mr. Levin ', 3);

INSERT INTO Courses
Values ('D1000', 3, 'Database Basics', '1:00pm', '3:00pm', '2/3/03',
'5/3/03', 'Mr. Carter', 3);

INSERT INTO Courses
Values ('A1000', 2, 'Accounting I', '2:00pm', '4:00pm', '2/3/03',
'5/3/03', 'Mrs. Smith', 3);

INSERT INTO Courses
Values ('A1001', 5, 'Accounting II', '1:00pm', '3:00pm', '2/3/03',
'5/3/03', 'Mrs. Terry', 3);

INSERT INTO Courses
Values ('P2000', 4, 'Physics', '2:00pm', '4:00pm', '2/3/03', '5/3/03',
'Mrs. Jones', 3);

INSERT INTO Courses
Values ('H1011', 1, 'Human Resource Mgt', '3:00pm', '5:00pm',
'2/3/03', '5/3/03', 'Mr. Pen', 3);

## Customers Table

CREATE TABLE Customers
(
CustomerID INTEGER NOT NULL PRIMARY KEY,
Firstname CHAR (50) NOT NULL,
Lastname CHAR (50) NOT NULL,
Address CHAR (50) NOT NULL,
City CHAR (20) NOT NULL,

*Appendix B*

State CHAR (2) NOT NULL,
Zipcode CHAR (10) NOT NULL,
Areacode CHAR (3) NULL,
PhoneNumber CHAR (8) NULL
);

## Insert Statements to Populate the Customers Table

INSERT INTO Customers
VALUES (1, 'Tom', 'Evans', '3000 2nd Ave S', 'Atlanta', 'GA', 98718, 301, '232-9000');

INSERT INTO Customers
VALUES (2, 'Larry', 'Genes', '1100 23rd Ave S', 'Tampa', 'FL', 33618, 813, '982-3455');

INSERT INTO Customers
VALUES (3, 'Sherry', 'Jones', '100 Free St S', 'Tampa', 'FL', 33618, 813, '890-4231');

INSERT INTO Customers
VALUES (4, 'April', 'Jones', '2110 10th St S', 'Santa Fe', 'NM', 88330, 505, '434-1111');

INSERT INTO Customers
VALUES (5, 'Jerry', 'Jones', '798 22nd Ave S', 'St. Pete', 'FL', 33711, 727, '327-3323');

INSERT INTO Customers
VALUES (6, 'John', 'Little', '1500 Upside Loop N', 'St. Pete', 'FL', 33711, 727, '346-1234');

INSERT INTO Customers
VALUES (7, 'Gerry', 'Lexingtion', '5642 5th Ave S', 'Atlanta', 'GA', 98718, 301, '832-8912');

INSERT INTO Customers
VALUES (8, 'Henry', 'Denver', '8790 8th St N', 'Holloman', 'NM', 88330, 505, '423-8900');

INSERT INTO Customers
VALUES (9, 'Nancy', 'Kinn', '4000 22nd St S', 'Atlanta', 'GA', 98718, 301, '879-2345');

INSERT INTO Customers
VALUES (10, 'Derick', 'Penns', '2609 15th Ave N', 'Tampa', 'FL', 33611, 813, '346-1232');

## Customers2 Table

CREATE TABLE Customers2
(
CustomerID INTEGER NOT NULL PRIMARY KEY,
Firstname CHAR (50) NOT NULL,
Lastname CHAR (50) NOT NULL,
Address CHAR (50) NOT NULL,
City CHAR (20) NOT NULL,
State CHAR (2) NOT NULL,
Zipcode CHAR (10) NOT NULL,
Areacode CHAR (3) NULL,
PhoneNumber CHAR (8) NULL
);

## Insert Statements to Populate the Customers2 Table

INSERT INTO Customers2
VALUES (1, 'Tom', 'Evans', '3000 2nd Ave S', 'Atlanta', 'GA', 98718, 301, '232-9000');

*Appendix B*

INSERT INTO Customers2
VALUES (2, 'Larry', 'Genes', '1100 23rd Ave S', 'Tampa', 'FL', 33618, 813, '982-3455');

INSERT INTO Customers2
VALUES (3, 'Sherry', 'Jones', '100 Free St S', 'Tampa', 'FL', 33618, 813, '890-4231');

INSERT INTO Customers2
VALUES (4, 'April', 'Jones', '2110 10th St S', 'Santa Fe', 'NM', 88330, 505, '434-1111');

INSERT INTO Customers2
VALUES (5, 'Jerry', 'Jones', '798 22nd Ave S', 'St. Pete', 'FL', 33711, 727, '327-3323');

INSERT INTO Customers2
VALUES (6, 'John', 'Little', '1500 Upside Loop N', 'St. Pete', 'FL', 33711, 727, '346-1234');

INSERT INTO Customers2
VALUES (7, 'Gerry', 'Lexingtion', '5642 5th Ave S', 'Atlanta', 'GA', 98718, 301, '832-8912');

INSERT INTO Customers2
VALUES (8, 'Henry', 'Denver', '8790 8th St N', 'Holloman', 'NM', 88330, 505, '423-8900');

INSERT INTO Customers2
VALUES (9, 'Nancy', 'Kinn', '4000 22nd St S', 'Atlanta', 'GA', 98718, 301, '879-2345');

INSERT INTO Customers2
VALUES (10, 'Derick', 'Penns', '2609 15th Ave N', 'Tampa', 'FL', 33611, 813, '346-1232');

SQL Script for the Tables Used in the Book

INSERT INTO Customers2
VALUES (11, 'Adam', 'Williams', '1333 5th St N', 'Tampa', 'FL', 33611, 813, '326-7777');

INSERT INTO Customers2
VALUES (12, 'Stan', 'Willows', '1837 30th Ave S', 'Tampa', 'FL', 33611, 813, '346-1100');

INSERT INTO Customers2
VALUES (13, 'Ricky', 'Canton', '1009 50th Ave N', 'Tampa', 'FL', 33611, 813, '346-3223');

INSERT INTO Customers2
VALUES (14, 'Pete', 'West', '2000 4th Ave N', 'Tampa', 'FL', 33611, 813, '346-8778');

## Employees Table

CREATE TABLE Employees
(
SocialSecNum CHAR (11) NOT NULL PRIMARY KEY,
Firstname CHAR (50) NOT NULL,
Lastname CHAR (50) NOT NULL,
Address CHAR (50) NOT NULL,
Zipcode CHAR (10) NOT NULL,
Areacode CHAR (3) NULL,
PhoneNumber CHAR (8) NULL
);

## Insert Statements to Populate the Employees Table

INSERT INTO Employees
VALUES ('266-73-1982', 'John', 'Dentins', '2211 22nd Ave N Atlanta, GA', 98718, 301, '897-4321');

*Appendix B*

INSERT INTO Employees
VALUES ('266-11-4444', 'Sam', 'Elliot', '1601 Center Loop Tampa, FL', 33612, 813, '898-2134');

INSERT INTO Employees
VALUES ('263-73-1442', 'Adam', 'Williams', '1938 32nd Ave S. St. Pete, FL', 33711, 727, '321-2234');

INSERT INTO Employees
VALUES ('226-73-1919', 'Jacob', 'Lincoln', '2609 40th Ave S Honolulu, HI', 96820, 808, '423-4111');

INSERT INTO Employees
VALUES ('249-74-1682', 'Jackie', 'Fields', '2211 Peachtree St N Tampa, FL', 33612, 813, '827-2301');

INSERT INTO Employees
VALUES ('123-88-1982', 'Debra', 'Fields', '1934 16th Ave N Atlanta, GA', 98718, 301, '897-3245');

INSERT INTO Employees
VALUES ('211-73-1112', 'Tom', 'Jetson', '1311 2nd Ave E Atlanta, GA', 98718, 301, '897-9877');

INSERT INTO Employees
VALUES ('980-22-1982', 'Shawn', 'Lewis', '1601 4th Ave W Atlanta, GA', 98718, 301, '894-0987');

INSERT INTO Employees
VALUES ('982-24-3490', 'Yolanda', 'Brown', '1544 16th Ave W Atlanta, GA', 98718, 301, '892-1234');

INSERT INTO Employees
VALUES ('109-83-4765', 'Shaun', 'Rivers', '1548 6th Ave S Atlanta, GA', 98718, 301, '894-1973');

## Members Table

CREATE TABLE Members
(
MemberID INTEGER NOT NULL PRIMARY KEY,
Firstname CHAR (50) NOT NULL,
Lastname CHAR (50) NOT NULL,
Address CHAR (50) NOT NULL,
City CHAR (20) NOT NULL,
State CHAR (2) NOT NULL,
Zipcode CHAR (10) NOT NULL,
Areacode CHAR (3) NULL,
PhoneNumber CHAR (8) NULL
);

## Insert Statements to Populate the Members Table

INSERT INTO Members
VALUES (1, 'Jeffrey', 'Lindley', '3980 14th Ave S', 'Atlanta', 'GA', 98700, 301, '451-5451');

INSERT INTO Members
VALUES (2, 'Jerry', 'Lindsey', '4000 3rd Ave S', 'Tampa', 'FL', 33600, 813, '923-7852');

INSERT INTO Members
VALUES (3, 'Gerry', 'Pitts', '3090 13th St N', 'Tampa', 'FL', 33611, 813, '286-4821');

INSERT INTO Members
VALUES (4, 'Stan', 'Benson', '1825 8th St N', 'Santa Fe', 'NM', 88388, 505, '464-1578');

*Appendix B*

INSERT INTO Members
VALUES (5, 'Peter', 'Gable', '1097 10th Ave S', 'St. Petersburg', 'FL', 33754, 727, '327-1253');

## Numbers Table

CREATE TABLE Numbers
(
Column1 INTEGER NOT NULL,
Column2 INTEGER NOT NULL,
Column3 INTEGER NOT NULL
);

## Insert Statements to Populate the Numbers Table

INSERT INTO Numbers
VALUES (20.00, 4, 21.3);

INSERT INTO Numbers
VALUES (10.00, 5, 20.3);

INSERT INTO Numbers
VALUES (30.00, 10, 16.8);

INSERT INTO Numbers
VALUES (50.00, 2, 18.3);

INSERT INTO Numbers
VALUES (60.00, 30, 12.6);

INSERT INTO Numbers
VALUES (70.00, 2, 2.1);

INSERT INTO Numbers
VALUES (10.00, 39, 2.9);

INSERT INTO Numbers
VALUES (40.00, 29, 19.2);

INSERT INTO Numbers
VALUES (80.00, 54, 15.8);

INSERT INTO Numbers
VALUES (20.00, 66, 23.1);

## Sales Table

CREATE TABLE Sales
(
SalesID INTEGER NOT NULL PRIMARY KEY,
SupplyID CHAR (7) NOT NULL,
CustomerID INTEGER NOT NULL,
DateSold DATETIME NOT NULL
);

## Insert Statements to Populate the Sales Table

INSERT INTO Sales
VALUES (1, 'AR100', 2, '2/3/03');

INSERT INTO Sales
VALUES (2, 'WC100', 8, '2/5/03');

INSERT INTO Sales
VALUES (3, 'AR100', 7, '2/6/03');

INSERT INTO Sales
VALUES (4, 'FL100', 1, '2/8/03');

INSERT INTO Sales
VALUES (5, 'MT100', 3, '2/8/03');

*Appendix B*

INSERT INTO Sales
VALUES (6, 'GR100', 4, '2/10/03');

INSERT INTO Sales
VALUES (7, 'WC100', 5, '2/22/03');

INSERT INTO Sales
VALUES (8, 'PS100', 9, '2/20/03');

INSERT INTO Sales
VALUES (9, 'CD100', 6, '2/18/03');

INSERT INTO Sales
VALUES (10, 'CP100', 10, '2/17/03');

INSERT INTO Sales
VALUES (11, 'CP100', 10, '2/17/03');

INSERT INTO Sales
VALUES (12, 'CP100', 5, '2/17/03');

INSERT INTO Sales
VALUES (13, 'CC100', 4, '2/17/03');

INSERT INTO Sales
VALUES (14, 'GR100', 3, '2/8/03');

INSERT INTO Sales
VALUES (15, 'MT100', 2, '2/17/03');

INSERT INTO Sales
VALUES (16, 'WC100', 1, '2/8/03');

INSERT INTO Sales
VALUES (17, 'CP100', 3, '2/8/03');

*SQL Script for the Tables Used in the Book*

## Supplies Table

CREATE TABLE Supplies
(
SupplyID CHAR (7) NOT NULL PRIMARY KEY,
SupplyName CHAR (50) NOT NULL,
Price MONEY NOT NULL,
SalePrice MONEY NOT NULL,
InStock INTEGER NOT NULL,
OnOrder INTEGER NOT NULL
);

For Microsoft SQL Server, substitute DECIMAL (8, 2) for the MONEY datatype.

## Insert Statements to Populate the Supplies Table

INSERT INTO Supplies
VALUES ('CD100', 'China Doll', 20.00, 18.00, 200, 0);

INSERT INTO Supplies
VALUES ('CP100', 'China Puppy', 15.00, 13.50, 20, 40);

INSERT INTO Supplies
VALUES ('WC100', 'Wooden Clock', 11.00, 9.90, 100, 0);

INSERT INTO Supplies
VALUES ('GR100', 'Glass Rabbit', 50.00, 45.00, 50, 20);

INSERT INTO Supplies
VALUES ('CC100', 'Crystal Cat', 75.00, 67.50, 60, 20);

INSERT INTO Supplies
VALUES ('PS100', 'Praying Statue', 25.00, 22.50, 3, 40);

*Appendix B*

INSERT INTO Supplies
VALUES ('MT100', 'Miniature Train Set', 60.00, 54.00, 1, 30);

INSERT INTO Supplies
VALUES ('DB100', 'Dancing Bird', 10.00, 9.00, 10, 20);

INSERT INTO Supplies
VALUES ('FL100', 'Friendly Lion', 14.00, 12.60, 0, 30);

INSERT INTO Supplies
VALUES ('AR100', 'Animated Rainbow', 20.00, 18.00, 10, 20);

# Appendix C

# SQL Command Syntax

This appendix provides the SQL syntax for some of the most commonly used SQL statements. Since SQL syntax varies slightly from one DBMS to another always refer to your DBMS documentation.

## Close Cursor

CLOSE CursorName

## CREATE INDEX

CREATE INDEX IndexName
ON Tablename (ColumnName, ColumnName)

## CREATE PROCEDURE

CREATE PROCEDURE ProcedureName [Parameters] [Optional]
AS
SQL QUERY

## CREATE TABLE

CREATE TABLE TableName
(
ColumnOne Datatype [NULL | NOT NULL], [Constraint],
ColumnTwo Datatype [NULL | NOT NULL], [Constraint],
ColumnThree Datatype [NULL | NOT NULL], [Constraint]
)

*Appendix C*

# CREATE TRIGGER

CREATE TRIGGER TriggerName
ON TableName
FOR INSERT | UPDATE | DELETE
AS Operation

# CREATE VIEW

CREATE VIEW ViewName AS
SELECT ColumnOne, ColumnTwo, ColumnThree
FROM TableName

# DEALLOCATE Cursor

DEALLOCATE CursorName

# DELETE Index

DROP INDEX TableName.IndexName

# DELETE Stored Procedure

DROP PROCEDURE ProcedureName

# DELETE Table

DROP TABLE TableName

# DELETE Trigger

DROP TRIGGER TriggerName

## DELETE VIEW

DROP VIEW ViewName

## INSERT Statement

INSERT INTO TableName [(ColumnNames, …)]
VALUES (values, …)

## Qualification

TableName.ColumnName

## SELECT Statement

SELECT ColumnName, ColumnName, ColumnName
FROM TableName

## WHERE Clause

WHERE [Condition]

# Appendix D

# Instructions on Where to Type SQL Script in Microsoft Access and Microsoft SQL Server

The examples throughout the book were created using Microsoft Access and SQL Server. This appendix contains instructions on how to locate the interface in which to type SQL script in Microsoft Access and Microsoft SQL Server.

## Microsoft Access

1. After launching Microsoft Access, either open an existing database or create a new database. If you create a new database, you must create and save a name for your database.

2. After you either create or open your database, click **Queries** on the left, and then click the "New" button located near the top of the screen.

3. Select Design View, and click OK.

4. You'll see a Show Table dialog box. Click close on this dialog box without selecting any tables.

5. Select the View button near the top of the screen.

 When you place your cursor over a button, the name of the button is displayed.

*Instructions on Where to Type SQL Script in Microsoft Access and Microsoft SQL Server*

6. Use the "View" button to select the SQL View. (Click the down arrow located on the "View" button to locate the SQL View).

7. Type your SQL commands in this view (SQL View).

To run a command, click the "Run" button. The run button is a red explanation mark.

## Microsoft SQL Server 7.0

1. Launch the Enterprise Manager application. Enterprise Manager is one of many applications that comes with Microsoft SQL Server 7.

If you are not connected to a server and you already created a database in Microsoft SQL Server, select the database by clicking on it to highlight it before you select the SQL Query Analyzer in step two below. This will cause the SQL Query Analyzer to open without prompting you to log in.

2. Click "Tools" from above and select SQL Query Analyzer.

3. Log into your server.

4. Select your database from the DB drop down box.

5. Type your SQL commands in the window.

6. To run a command, click the "Execute Query" button or the F5 button on your keyboard.

# Index

## @

@@ERROR, 147
@@FETCH_STATUS, 158

## A

Add, 131, 139
Aggregate functions, 56, 58, 74
All, 100, 101, 107
Alter, 131
Alter table, 139
And, 40, 45, 46
Any, 100, 101, 105
Arithmetic functions, 57
Arithmetic operators, 52, 54
AS, 25, 31
Asc, 66, 69
Asterisk, 8, 21, 72
Avg (), 56, 58

## B

Begin transaction, 143, 145
Between, 40, 49, 50

## C

Calculated fields, 52
Cartesian product, 81, 83
Character functions, 57
Character operators, 39
Check, 134, 135
Clause, 66
Close, 143, 158
Clustered index, 131
Column, 1
Commit transaction, 143, 145
Comparison operators, 39
Composite index, 131, 137

Concatenation, 25, 32, 52, 53
Constraint, 131
Count (), 56, 58, 75, 114
Create index, 131, 137, 138
Create procedure, 143, 148
Create table, 8, 10, 133
Create table syntax, 9
Create trigger, 143, 156
Create view, 117, 118
Cursor, 143, 158

## D

Datatype, 8, 13
Date and time functions, 57, 61
Deallocate, 143
Declare, 158
Delete, 8, 23
Delete an index, 139
Desc, 66, 70
Descending order, 70
Direct Access Method, 131, 136
Distinct, 25, 29
Drop column, 131, 139, 141
Drop index, 132
Drop table, 132, 141
Drop trigger, 158
Drop view, 117

## E

Execute, 144
Exists, 100, 101, 104
Expression, 38

## F

Fetch, 144, 158
Fetch next from, 144, 160
Field, 8
Field size, 8

*Index*

Foreign key, 1, 3, 134, 136
From, 9, 25, 26
Functions, 52, 56

## G

Global variables, 158
Group by, 66, 74

## H

Having, 66, 77

## I

IF, 144
IF statement, 147
IN, 40, 100, 101, 102, 103
Index, 132, 137
Inner join, 81, 83
Input parameter, 153
Insert, 17
Insert Into, 17
Insert statement, 9, 16
Insert syntax, 16

## J

Joins, 81, 82

## K

Keys, 1, 3, 81
Keyword, 1, 6, 10

## L

Left outer join, 81, 92
Like, 39, 42
Logical operators, 40, 45

## M

Max (), 56, 58, 113

Microsoft Access, 13
Microsoft SQL Server, 14
Min (), 56, 58

## N

Natural join, 81, 89
Next, 144, 158
Normalization, 1, 4
Not, 40, 46
Not null, 9, 11, 134
Null, 9, 11
Numeric datatype, 135

## O

ON, 81, 93, 156
Open, 144
Open cursor, 158
Operators, 38
OR, 40, 45, 46
Order by, 66, 67
Outer join, 82, 92

## P

Parameters, 148
Primary key, 2, 3, 134
Print, 144

## Q

Qualification, 82
Query, 2, 100

## R

Referential integrity, 132
Relational database, 2
Replace (), 63
Result set, 144, 158
Right outer join, 82, 92, 96
Rollback, 144
Rollback transaction, 145
Row, 2
Rtrim (), 64

## Index

## S

Save transaction, 144, 145
Savepoint, 144, 145
Select, 9, 25, 26
Select statement, 25, 26
Select statement syntax, 26
Self-join, 82, 86
Sequential Access Method, 132, 136
Set, 9, 117
SQL, 6
Stored procedure, 144, 148
Structured Query Language, 2, 5
Subquery, 100, 101
Sum (), 56, 58
Syntax, 2, 6

## T

Table, 2
Table creation, 8

Transaction, 144, 145, 146
Transaction processing, 144, 145
Trigger, 144, 155

## U

Underscore, 40
Union, 25, 33, 34
Union all, 25, 33, 35
Unique, 134
Update, 9, 21, 117, 125

## V

View, 117, 118

## W

Where, 9, 22, 38
Where clause syntax, 38

Printed in the United States
68677LVS00004B/73